German C-Types Volume 1

Jack Herris

German C-Types Volume 1

Jack Herris

Great War Aviation Centennial Series #88

Acknowledgements

Brandenburg KDD, C.II, and L 15 text courtesy of Colin Owers. Thanks to Colin Owers, Bruno Schmäling, Michael Schmeele, Greg VanWyngarden, and Reinhard Zankl for photographs and information.

Color aircraft profiles © Bob Pearson. Bob notes that while every care is taken, some colors are speculative based on known practices at the time. Purchase Bob's CD of WWI aircraft profiles for $50 US/Canadian, 40 €, or £30, airmail postage included, via Paypal to Bob.

For information on our aviation books, please see our website at: **www.aeronautbooks.com**. Aeronaut Books is looking for photographs of rare German aircraft of WWI for our books. To help please contact the publisher at **jherris@me.com**.

Interested in WWI aviation? Join The League of WWI Aviation Historians (**www.overthefront.com**), Cross & Cockade International (**www.crossandcockade.com**), and Das Propellerblatt (**www.propellerblatt.de**).

ISBN: 978-1-964637-02-0

Design and layout: Jack Herris
Cover design: Jack Herris
Color Profiles: Bob Pearson; bpearson@kaien.net
Digital photo editing: Jack Herris

Books for Enthusiasts by Enthusiasts

www.aeronautbooks.com

Table of Contents

Volume 1

Introduction 4
AEG 8
Ago 22
Albatros 46
Aviatik 109
Brandenburg 134
DFW 146
Euler 170

Volume 2

Fokker 4
Friedrichshafen 7
Germania 8
Halberstadt 20
Hannover 40
Kondor 42
LVG 44
Mercur 92
Otto 94
Pfalz 104
Roland 105
Rumpler 121
Schütte-Lanz 164
Zeppelin-Lindau 165

Preface

The two volumes of German C-Types are an encyclodia of armed two-seat reconnaissance planes of WWI. All material herein has been published in previous volumes on airplanes by manufacturer.

Reconnaissance was the most important aviation role of WWI; reflecting that, 50% of German warplanes were C-types. Because aviation technology of the time was primitive, the airplanes having limited speed, range, and payload, bombing and strafing were not decisive. Although fighters have received the greatest attention, their true purpose was to stop the reconnaissance planes and bombers.

Introduction

These two volumes summarize German C-Types, which are armed, two-seat reconnaissance airplanes. Almost all tasks aviators have done were first tried during WWI. However, the primitive aviation technology of the time ensured that reconnaissance was the most successful and valuable role. By WW2, improved aviation technology gave aircraft the range, speed, and payload to make bombing effective and the most important role of aviation, although reconnaissance was still crucial and remains so to this day.

The composition of the German air force reflected the importance of reconnaissance. Although fighters have received the most popular coverage and attention, they comprised only 42% of German warplanes at the front by 1918. In contrast, reconnaissance aircraft comprised 50% of aircraft at the front. By 1918 these reconnaissance were armed biplanes, called C-Types by Germany.

To perform their mission, whether short-range or long-range reconnaissance or artillery spotting, C-Type crews had to fly over Allied lines, where they were subject to anti-aircraft fire and fighter interception. In contrast, Germany was on the strategic defensive due to its permanent numerical inferiority. As a result, most German fighter activity was over German lines where their mission was to intercept Allied aircraft.

One benefit of being over your own lines during combat was the ability to dive out of the fight when you are in a bad position. Another was the possibility of returning to combat after being shot down instead of being captured.

Due to the precision machining necessary, aero engines take a lot of time and resources to develop and bring to production. On the other hand, airframes of the period were relatively simple and easy to design and build. One result was Germany's focus on continuously developing new airframes to maintain competitive performance. This meant that German aircraft manufacturers designed and built many new C-Types.

When WWI started, the available airplanes were very primitive and none of the air services had air combat experience. The available aircraft were also unarmed. Germany entered the war with unarmed monoplanes (A-Types) and biplanes (B-Types). Engines for these aircraft generally offered 80–120 hp.

Perhaps the easiest way to undertand the story of German reconnaissance airplanes is to think of them in terms of generations, primarily based on the available engines.

First Generation C-types

As air combat developed, the need for defensive armament became obvious. The biplanes of the time were more robust than monoplanes, and the first C-Types (armed two-seat biplanes) were developed from B-Types by using more powerful engines, typically 150–160 hp, to produce adequate performance despite the extra weight of the machine gun(s) and ammunition.

- AEG C.I, C.IV
- Ago C.I (pusher)
- Albatros C.I, C.III, C.IV, C.VI
- Aviatik C.I, C.III
- Brandenburg C.I (Austro-Hungarian use only)
- DFW C.I, C.II, C.IV
- Euler C.I (training use only)
- Friedrichshafen C.I (prototype only)
- Germania KDD (prototype only)
- LVG C.I, C.II
- Otto C.I (pusher), C.II (training use only)
- Roland C.II
- Rumpler C.I

The most successful and numerous B-types were built by Albatros, Aviatik, and LVG. Unsurprisingly, the most numerous first-generation C-Types were built by these same companies. The observers in the Albatros and LVG C-Types were moved to the rear cockpit to give them the largest field of fire; Aviatik C.I retained the observer in the front cockpit to preserve the best field of view despite the more limited field of fire. Unsurprisingly, experience soon showed the armed observer needed to be in the rear cockpit to provide the best protection against Allied fighters.

Second Generation C-types

Availability of the 200 hp Benz Bz.IV engine enabled the second generation of C-Types.

- Ago C.IV, C.VI, C.VI, C.VII
- Albatros C.VII
- Aviatik C.II, C.IV, C.VI, C.IX
- DFW C.V, C.VI, C.VII
- Halberstadt C.III, C.V
- LVG C.V, C.VI
- Rumpler C.III

The robust and reliable DFW C.V, with its excellent handling qualities, became the most successful of these. Built by DFW and other companies under license, it became the most numerous German warplane of WWI. Introduced in late 1916, the DFW C.V served at the front until the Armistice despite its modest speed.

The Ago C.I was one of the very few German pusher designs. It was sturdy and easy to fly, but slow. The Ago C.IV was intended to be its faster replacement. The C.IV was the fastest C-Type of the time, but was fragile and had very poor handling qualities. After many accidents, the C.IV was modified to solve these issues. By the time the modified C.IV reached the front, its reputation was so bad crews refused to fly it and it was removed from the front a final time.

The LVG company wanted to make more profit by

building its own designs instead of building the DFW C.V under license, so hired the designer of the DFW C.V to produce the LVG C.V, essentially an improved DFW C.V. The C.VI was a further refinement and replaced the C.V in production. The LVG C.VIII was derived from the C.VI but was too late for combat.

Halberstadt is well-known for its CL.II and CL.IV two-seat fighters, but its C.V, introduced in the summer of 1918, was the best short-range C-Type to see service. The C.V was not successful in long-range missions; as a result the C.VIII, using a high-altitude engine, was developed for long-range missions but was too late for production.

Long-Range Second Generation C-types

Second generation C-types were generally powered by the 200 hp Benz Bz.IV. However, availability of the 220 hp Mercedes D.IV, a straight-8 engine created by adding two more cylinders to the 160 hp Mercedes D.III, enabled creation of several special types optimized for long-range reconnaissance due to their relatively high speed.

- AEG C.V (prototype only)
- Albatros C.V
- Kondor W 2C (prototype only)
- LVG C.IV

The Albatros C.V and LVG C.IV were placed in production. Upon reaching the front, they were as fast as Allied fighters and served successfully due to that. They were eventually withdrawn from service when faster Allied fighters appeared.

Third Generation C-types

Availability of the powerful 260 hp Mercedes D.IVa engine enabled creation of a third generation of C-types.

- Ago C.IX (prototype only)
- Albatros C.X, C.XII
- Halberstadt C.VII, C.VIII (prototypes only)
- Pfalz C.I (license-built Rumpler C.IV)
- Rumpler C.IV (best long-range type)

While the DFW C.V and its LVG derivatives were successful short-range types, Rumpler developed the best German long-range reconnaissance types. Despite its very limited quantity of 49, the Rumpler C.III was produced in at least three configurations as Rumpler worked to refine its airframe. The 200 hp Benz of the final C.III airframe was replaced by the 260 hp Mercedes to create the C.IV. The use of different engines and equipment created a family of types optimized for different missions. The C.IV and its variants had excellent ceiling and speed and were very difficult to intercept at high altitude. Its planned replacement, the Rumpler C.X, was too late to see operational service.

Albatros was successful into 1917 with its C.III and C.V, and its C.VII was produced in quantity but was not the best C-Type of its time. Subsequent Albatros C-type designs were mediocre performers despite their appealing lines and had little impact.

Fourth Generation C-types

The fourth and final generation of C-types was characterized by advanced technology, either advanced airframes, such as all metal airframes, or advanced engines (V-type).

- Aviatik F.I (V12 engine)
- Rumpler C.X (advanced airframe)
- Zeppelin-Lindau C.II (all-metal airframes)

The final long-range reconnaissance design to reach prototype status before the Armistice was the Rumpler C.X. It had a high-altitude Maybach Mb.IVa six-cylinder engine and refined, advanced airframe designed to replace the C.IV family aircraft. It flew extensively as a prototype but was too late to see production before the war ended. The Zeppelin-Lindau all-metal biplanes were placed into production too late to see combat service but the C.II saw successful postwar service in Switzerland.

The Aviatik F.I (designed to the new 'F' category for *Fernaufklärerflugzeuge* or long-range aerial reconnaissance aircraft defined by *Idflieg* in September 1918) first flew in December 1918. Since this was after the Armistice, it was completed as a transport. It was powered by the 350 hp Benz Bz.V V-12 engine that was also too late for the war.

In 1914 *Idflieg* stopped development of V-engines in favor of inline six-cylinder engines with their lower frontal area. This edict remained in place until *Idflieg* evaluated the Hispano-Suiza V-8 from a captured Spad 7 in late 1916. The Hispano-Suiza enginen had twice the power for its weight as the popular Mercedes D.III, although the Mercedes was far more reliable and durable. As a result, in December 1916 *Idflieg* requested development of a comparable V-8 fighter engine from German industry. Due to limited time and resources, the Benz and Daimler V-8 engines were not certified for production until very late in the war; too late to power any warplane in combat. Similarly, despite development of prototype V-10, V-12, and W-18 engines, none were ordered into production until they were too late to appear in combat. The result of *Idflieg's* myopia was that no German warplane powered by a V-8 or V-12 appeared in combat during the war.

Above: The Rumpler C.X was the ultimate long-range reconnaissance plane; it was too late for production.

Above: The Aviatik C.I was a successful, first generation C-type. Later Aviatik ended up building the DFW C.V.

Above: The Albatros C.I was a successful, first generation C-type.

Above: The DFW C.V was the most numerous German warplane; introduced in late 1916 it served until the Armistice.

Above: The Ago C.IV was a fast reconnaissance type removed from the front due to its high accident rate from very poor handling qualities and its fragility.

Above: The Halberstadt C.V was the best German short-range reconnaissance two-seater; the long wingspan is distinctive.

Above: The LVG C.V was a refined development of the DFW C.V, being designed by the same designer.

Above: The Rumpler C.IV was Germany's best high-altitude, long-range reconnaissance plane.

C-Type Front-Line Inventory

Manufacturer and Type		1914			1915						1916						1917						1918			
		31 Aug	31 Oct	31 Dec	28 Feb	30 Apr	30 Jun	31 Aug	31 Oct	31 Dec	28 Feb	30 Apr	30 Jun	31 Aug	31 Oct	31 Dec	28 Feb	30 Apr	30 Jun	31 Aug	31 Oct	31 Dec	28 Feb	30 Apr	30 Jun	31 Aug
AEG	C.I						4	10	28	26	34	34		1	1	1	1	1								
	C.II											1														
	C.IV								1						9	42	90	124	127	63	88	68	40	26	29	34
Ago	C.I						5	5	9	14	22	15	23	16	9	8	5	1								
	C.II									2	1															
	C.III										1															
	C.IV															8	25	1	19	90	42	13	3			
Albatros	C (Hirth)			1																						
	C.I					1	36	91	184	278	349	213	165	143	125	90	52	37	46	13	4	2		1		1
	C.III									12	51	208	300	354	320	320	214	174	107	29	15	5	1	1	2	
	C.IV											1					1	2		2						
	C.V													65	57	47	50	60	57	29	8	5	5			
	C.VI													8	71	53	111	21	3	1						
	C.VII											2			94	249	372	296	165	74	33	11	6	4		1
	C.VIII																	1	1	1	1					
	C.IX																			2	2	1	1	1		
	C.X														2			70	98	47	26	11	4			
	C.XII																		41	93	92	66	48	36	20	15
	C Lf																3	1								
Aviatik	C.I					1	8	9	35	81	161	150	219	183	177	110	63	27	12	1	1					
	C.II								3		7	6				22	62	50	34	4	1					
	C.III													26	47	22	20	6	2							
	C.V																		1							
DFW	C.I									25			2	2	9	6	5	3	1	1						
	C.V														21	42	79	341	839	1057	901	845	614	665	623	620
Gotha	C.I								2																	
Halberstadt	C.I																		4							
	C.V																		3						10	192
	C.IV																			15	65				40	
	C (exp.)																					1				
Hannover	C.I																		2							
	C.II																				19					
LVG	C.I						30	60	35	45	25	11	4	5	1	1		1								
	C.II						8	6	20	138	214	249	255	220	195	155	87	45	20	8	2	3				
	C.IV									1	1	1	1	27	60	76	58	39	14	1						
	C.V																			98	219	446	504	565	312	133
	C.VI																								173	400
Otto	C.I			1					1	9	1	1														
Pfalz	C.I			2																	3	24	11	7	9	6
	C.IV																		2							
Roland	C.I												35				21	9	1							
	C.II											17	35	37	45	40	53	42	1	1						
	C.IIa													13	11	24			12	1						
	C.III																1									
Rumpler	C.I									29	67	120	144	202	231	190	145	107	55	37	40	16	11	9	5	4
	C.Ia												1		2	1	1	52	162	129	30	5	3	3	2	1
	C.III															1	42	22	15	7	1	1	1	1		
	C.IV																	24	122	257	225	208	161	187	182	110
	C.V																				3	9			4	6
	C.VII																					17	48	94	112	85
	C.IX																					20	13	11	5	2
Unidentified																						20				
Total:				4		2	91	181	318	660	934	1029	1184	1302	1487	1508	1561	1557	1966	2061	1821	1797	1475	1611	1528	1610

AEG C.I

Above: AEG C.I 90/15 (above) wears black and white bands on its fuselage and carries white identification pennants. Powered by a 150 hp Benz Bz.III like the AEG B.II, the AEG C.I still retained the side radiators normally associated with B-types. About 70 AEG C.I reconnaissance two-seaters were built. The C.I below may be 90/15 but that is not confirmed.

The AEG C.I (company designation KZ9 for *Kampf-Zweidecker* 9, or battle biplane 9) was derived from the earlier, unarmed B.II by adding a machine-gun turret; there was no fixed gun for the pilot. Although structurally the same as the B.II, with steel-tube construction except for wooden wing ribs, the C.I had a different wing planform than the B.II; however, the wing-folding mechanism was retained. *Idflieg* ordered six C.I aircraft in early 1915 and the C.I prototype was completed in March. Additional C.I aircraft were ordered, a total of about 70 being delivered.

The AEG C.I was among the first C-type aircraft to reach the front, serving from June 1915 to April 1916. Climb rate was mediocre despite the C.I being lighter than most contemporary C-types. Due to its sturdy metal tube airframe the C.I was used well into 1917 as a trainer after being removed from the front.

AEG C-Type Production Orders		
Serial Numbers	**Qty**	**Order Date & Notes**
AEG C.I (70 Total)		
C.74–79/15	6	Unknown
C.82–93/15	12	April 1915
C.130–153/15	24	May 1915
C.324–348/15	25	May 1915
C.1366–1368/17	3	Unknown
AEG C.II (37 Total)		
C.869–888/15	20	September 1915
C.1887–1902/15	16	October 1915
C.9421/16	1	C.II prototype?
AEG C.IV (495 Total)		
C.1024–1123/16	100	March 1916
C.6575–6674/16	100	October 1916 (note 1)
C.1700–1799/17	100	April 1917
C.4800–4899/17	100	(May) 1917
C.7052–7126/17	75	July 1917 (note 2)
?	8	December 1917 (note 3)
C.1100–1111/18	12	February 1918 (note 3)
AEG C.IV(Fok) (200 Total)		
C.240–439/17	200	January 1917
AEG C.IVa(Fok) (100 Total)		
C.6500–6699/17	100	January 1917 (note 4)

Notes:
1. Initially order was for 100 Rumpler C.IV(AEG)
2. 25 of these for Turkey, 50 for Germany
3. Special equipment for tropics
4. 200 ordered but only 100 built

AEG C.II & C.IIa – C.IIe

The AEG C.II was basically a lightened C.I with a modified wing planform. Like the C.I it was powered by the 150 hp Benz Bz.III engine. It was 122 kg lighter than the C.I, giving it better performance, but airframe strength was compromised, resulting in a number of failures during load tests. Modifications to strengthen the airframe took time and the first six C.II aircraft were not delivered until April 1916. One C.II was sent to the front that month, but was soon removed from the front and sent to join the other C.II aircraft for training use.

From early 1915 through the spring of 1916, AEG modified C.II biplanes with different airfoil sections, chord, span, and gap, with the different configurations given factory (not *Idflieg*) designations C.IIa through C.IIe. Data was collected that led to aerodynamic improvements to wing and radiator design, some of which were key in the design of the AEG C.IV.

The AEG C.IIe tested an armored, annular radiator (Patent 299,725) in which the cooling air was forced through the radiator by a fan attached to the propeller. This particular development was driven by combat experience with the armored AEG J.I, which demonstrated that the radiator was the most vulnerable part of the armored aircraft during low-level ground attacks.

Above: AEG C.II 885/15 was from the first C.II production batch.

Right: AEG C.II, likely 885.15. The C.II retained the primitive side radiators used in the AEG B-types.

Above: AEG C.I 75/15 was powered by a 150 hp Benz Bz.III and still used side radiators.

AEG C-Type Specifications					
	C.I	C.II	C.III	C.IV	C.IV(Fok)
Engine	150 hp Benz Bz.III	150 hp Benz Bz.III	150 hp Benz Bz.III	160 hp Mercedes D.III	160 hp Mercedes D.III
Span Upper	13.07 m	11.95 m	12.00 m	13.00 m	12.99 m
Span Lower	11.83 m	11.45 m	11.30 m	12.50 m	12.36 m
Chord Upper	1.60 m	1.85 m	1.60 m	1.65 m	1.65 m
Chord Lower	1.60 m	1.85 m	1.60 m	1.65 m	1.65 m
Gap	1.95 m	1.80 m	1.40 m	1.95 m	1.95 m
Wing Area	36.0 m^2	39.0 m^2	33.90 m^2	39.0 m^2	—
Length	7.95 m	7.09 m	6.50 m	7.20 m	7.22 m
Track	2.30 m	2.15 m	2.15 m	2.15 m	2.16 m
Empty Weight	710 kg	680 kg	687 kg	800 kg	901 kg
Loaded Weight	1,125 kg	1,200 kg	1,232 kg	1,320 kg	1,414 kg
Maximum Speed	130 kmh	138 kmh	145 kmh	158 kmh	—
Climb, 1000m	11 min.	8 min.	—	6 min.	4.8 min.
Climb, 2000m	25 min.	18 min.	—	12.5 min.	12.1 min.
Climb, 3000m	55 min.	41 min.	—	23 min.	21.9 min.
Climb, 4000m	—	—	—	38 min.	35.4 min.
Armament	1 flexible machine gun, small bombs	1 flexible machine gun, 4x10 kg bombs	1 flexible machine gun	1 fixed & 1 flexible machine gun	1 flexible machine gun

Above: AEG C.I 139/15 has broken its back in a hard landing. It wears national insignia on its fuselage as well as its rudder.

AEG C-Type Specifications				
	C.V	C.VII	C.VIII	C.VIIIDr
Engine	220 hp Mercedes D.IV	160 hp Mercedes D.III	160 hp Mercedes D.III	160 hp Mercedes D.III
Span Upper	13.26 m	11.10 m	9.50 m	11.20 m
Span Lower	12.45 m	10.05 m	9.10 m	10.40 m
Chord Upper	1.75 m	1.55 m	1.74 m	1.45 m
Chord Lower	1.75 m	1.30 m	1.33 m	0.82 m
Gap	2.07 m	1.85 m	1.60 m	1.00 m
Wing Area	41.5 m^2	26.0 m^2	22.67 m^2	31.0 m^2
Length	7.60 m	6.20 m	6.20 m	6.90 m
Track	2.30 m	2.00 m	2.10 m	1.90 m
Empty Weight	900 kg	758 kg	800 kg	800 kg
Loaded Weight	1,432 kg	1,118 kg	1,160 kg	1,160 kg
Maximum Speed	165 kmh	175 kmh	170–190 kmh	158 kmh
Climb, 1000m	7 min.	4 min.	3.8 min.	3.8 min.
Climb, 2000m	13 min.	—	—	—
Climb, 3000m	22 min.	—	—	—
Climb, 4000m	37.5 min.	—	—	—
Armament	1 fixed & 1 flexible machine gun	1 fixed & 1 flexible machine gun	1 fixed & 1 flexible machine gun	1 fixed & 1 flexible machine gun
Note: C.VIII Triplane middle wing had 10.80 m span, 0.82 m chord. Climb to 5,000 m in 34 min.				

Above: The AEG C.IIe tested a patented armored, annular radiator developed for the armored J-types.

Below: AEG C.II with an additional gun mounted over the wing. The C.II did not have a synchronized gun for the pilot.

AEG C.III

Above: Like the earlier C.I and C.II, the AEG C.III was powered by a 150 hp Benz Bz.III. (The Peter M. Bowers Collection/The Museum of Flight)

The AEG C.III was designed to give the observer the maximum field of fire, achieved by placing the top wing at the lever of the top of the fuselage and the observer in the front cockpit. The Roland C.II to similar format, although with pilot in front, first flew in October 1915 and the AEG C.III, clearly inspired by the Roland, flew two months later.

The C.III was developed from the C.II and used the same 150 hp Benz Bz.III engine. One or two prototypes were built but the insufficient wing gap caused interference in the airflow between the upper and lower wings, limiting performance and flying qualities, and further development was abandoned in favor of more conventional configurations.

Above: The C.III was designed for a more offensive role than the C.I or C.II and the observer with his flexible gun was in the front cockpit, giving him a good field of fire, but it was not synchronized so could not safely fire through the propeller arc.

AEG C.IV

Above: AEG C.IV 1715/17 is run up before its next mission while the aircrew and ground crew pose for a team photograph. The C.IV, powered by the 160 hp Mercedes D.III, was the first AEG C-type to have a synchronized gun for the pilot.

The AEG C.II had been insufficiently robust for front-line service and could only be used for training, and the C.III did not have the performance of the Roland C.II. To develop a successful C-type AEG based its C.IV design on the results of the engineering tests done on the C.IIa through C.IIe development prototypes. Attention was also given to speeding production through simplifying the design, rationalizing assembly methods, and building welding jigs to enable production by semi-skilled labor. Outwardly similar to the previous C.I and C.II, the C.IV embodied many detailed structural and aerodynamic improvements and a new wing design that eliminated the unnecessary wing-folding mechanism. Due to the unfortunate experience of the C.II, the C.IV airframe was strengthened and used structural joints machines from solid billets instead of built-up from welded sections. As a result the C.IV had improved strength and durability compared to any of its AEG predecessors.

Powered by a 160 hp Mercedes D.III engine, the C.IV prototype first flew in March 1916. *Idflieg* awarded AEG a production contract for 100 C.IV aircraft in March 1916 pending successful completion of the static load tests. Unfortunately, a wing spar failed load testing in June and *Idflieg* refused to accept the AEG C.IV. *Idflieg* then suggested that AEG build the all-wood Rumpler C.IV under license, but this was clearly not suited for the AEG production methods geared to metal-framed aircraft and the proposal was quickly dropped. In the meantime the AEG C.IV had managed to pass its load tests and was approved for service, the first aircraft reaching the front in October 1916.

At the front the AEG C.IV soon gained a reputation for reliability and speed, but was not an easy aircraft to fly compared to the well-liked Albatros C.III and DFW C.V and some units did not appreciate it. In addition to reconnaissance it also had a useful bombing ability. Never famous like the shapely Roland C.II nor the exceptional Rumpler C.IV, the AEG C.IV was a solid, reliable aircraft that was capable enough to be developed into the production AEG N.I and AEG J.I and J.II.

Above: A crewman strikes a jaunty pose by AEG C.IV 1748/17 of the third production batch. The two-color sprayed camouflage typical of AEG practice at the time shows up nicely.

Below: AEG C.IV 7123/17 v.R. (*verlängerter Rumpf* = lengthened fuselage) serving postwar with the Polish air force. Large serial numbers were used with training aircraft.

Above: The AEG C.IV prototype (second version) benefitted from an improved steel-tube structure compared to the C.II.

Its robust metal airframe recommended it for service in the Middle-East where weather and temperature extremes were typical, and C.IV aircraft modified for the heat with larger radiators were delivered to both Turkish units and German units serving in the area. However, the aircraft's structure required welding equipment for even minor repairs, a factor that had been inexplicably overlooked when units using the type quickly found a nearly complete absence of the needed equipment in the area. This small but critical problem seriously reduced C.IV serviceability in the theater.

In January 1918 work was in progress to install a flexible 2cm Becker cannon in a C.IV modified with a larger observer's gun ring. The gun ring operated smoothly but the gunner's position was too cramped for proper loading and aiming of the cannon, and the project was cancelled in April.

AEG C.IV v.R.

AEG lengthened the C.IV fuselage to improve its flying and landing qualities; the modified aircraft was designated the AEG C.IV v.R. (*verlängerter Rumpf* – lengthened fuselage). This modification is not mentioned in any existing *Idflieg* records, although the AEG J.II was given a longer fuselage for the same reasons, as were some AEG J.I aircraft. To what extent the long fuselage was incorporated in late production aircraft is unknown.

AEG C.IV(Fok)

By the end of 1916 the Fokker company badly needed a suitable aircraft to manufacture because its own designs had been restricted to training service due to poor design and quality control. To maintain production output and the existing workforce, *Idflieg* contracted for license production of the AEG C.IV for training service. Fokker was the only other German company with the required welding skills for airframes, and in January 1917 received an order for 200 trainers as the AEG C.IV(Fok). Production deliveries began in July 1917 and the first production aircraft was type-tested in August 1917; deliveries were completed in December 1917.

AEG C.IVa(Fok)

Idflieg ordered 200 AEG C.IVa(Fok) trainers from Fokker in June 1917. The C.IVa(Fok) differed from the C.IV(Fok) in having a 180 hp Argus As.III engine in place of the 160 hp Mercedes D.III; the Mercedes was needed for Albatros and Pfalz fighter production and the Argus was readily available. Fokker delivered 100 of these aircraft between November 1917 and March 1918. However, the remaining 100 aircraft were cancelled to enable Fokker to use the production capacity for its new D.VII fighter.

AEG C.V

When a new engine became available it was common for *Idflieg* to order new prototypes from several manufacturers; this gave more experience with the new engine and enabled *Idflieg* to choose the best aircraft for production. When the new 220 hp Mercedes D.IV straight-eight became available *Idflieg* ordered new C-types powered by this engine from AEG, Albatros, and LVG. The resulting AEG C.V strongly resembled the AEG C.IV that was designed in parallel. Interestingly, the C.V was completed in February 1916, one month before the C.IV. Having a heavier engine, the C.V was slightly larger than the C.IV and was slightly faster. The competing Albatros C.V and LVG C.IV apparently had better performance or flying qualities than the AEG C.V because they were both ordered into production while the AEG C.V was not. Production of the Mercedes D.IV was limited and the engines allocated to AEG were used in the AEG G.III twin-engine bomber.

Above & Right: Despite its additional 60 hp, the AEG C.V was only slightly faster than the C.IV designed in parallel.

AEG C.VI

Apparently the AEG C.VI was an un-built project because there is no information about it in existing German records and no photographs exist of an AEG C.VI.

AEG C.VII

In August 1916 *Idflieg* issued a requirement for a two-seat escort aircraft. This was to be a lightened C-type, later categorized as a CL-type, powered by a 160–180 hp engine. AEG was one of the companies that responded with a proposal, and in October *Idflieg* gave AEG a contract for three prototype C.VII aircraft. The C.VII was powered by a 160 hp Mercedes D.III and the first aircraft was completed in December 1916. It was about 20% smaller than the C.IV powered by the same engine. Initial flight testing was apparently done in January 1917 at Nieder-Neuendorf, and the slightly modified C.VII prototype underwent testing in March. The initial *Idflieg* report (dated April 30, 1917) said the C.VII offered little that was new. The observer's field of fire was considered poor for an escort aircraft and the performance, while superior to the C.IV, was not as good as expected although flying qualities were acceptable. The C.VII was recommended for service on the less-demanding Eastern Front providing the load tests were passed. That recommendation was not compelling and no production was undertaken.

The third C.VII prototype was built with a highly-swept-back wing with minimum gap, intended to improve the gunner's field of fire. However, the competing Halberstadt and Hannover CL-types were superior and were placed in production while the AEG C.VII was not.

Above & Left: Either the first or second AEG C.VII prototype is shown here. Designed as an escort fighter, a category that later became the CL-class, the C.VII was acceptable but not superior, and the excellent Halberstadt CL.II and Hannover CL.II were produced instead.

Above & Below: The third AEG C.VII prototype featured a swept-back upper wing with minimum gap to improve the gunner's field of fire. However, the competing Halberstadt and Hannover CL-types were still superior aircraft.

AEG C.VIII & C.VIII Triplane

The AEG C.VIII was an attempt by AEG to improve the C.VII for the light two-seater escort role. Three prototypes were ordered, all powered by the 160 hp Mercedes D.III. Apparently two were completed as biplanes, the first in July 1917, and the third as a triplane. It should be remembered that the summer of 1917 was the height of *Idflieg's* triplane craze, which no doubt influenced AEG to complete one C.VIII as a triplane. AEG apparently hoped to achieve high speed by reducing drag to a minimum; the AEG history claims 190 km/h but engineering specifications show a top speed of 170 km/h. If the higher speed were actually achieved the C.VIII might have been ordered into production, but it was not and a reasonable conclusion is the 190 km/h top speed was not reached.

The C.VIII triplane used the engine, fuselage, and tail of the biplane C.VIII. Completed in October 1917 and flight tested in November, the C.VIII triplane climbed to 5000 meters in 34 minutes, a climb equivalent to the Albatros D.V single-seat fighter. With 80 kg less load it climbed to 5000 meters in 24 minutes. However, it was slower than the biplane C.VIII and flight-testing revealed the C.VIII triplane to have unsatisfactory flying qualities. AEG had proposed a smaller, lighter triplane in November but the disappointing flight-test results of the C.VIII triplane eliminated that idea as well.

Above & Facing Page, Top: One of the first two AEG C.VIII prototypes; the C.VIII was a cleaner, more aerodynamic design than the C.VII.

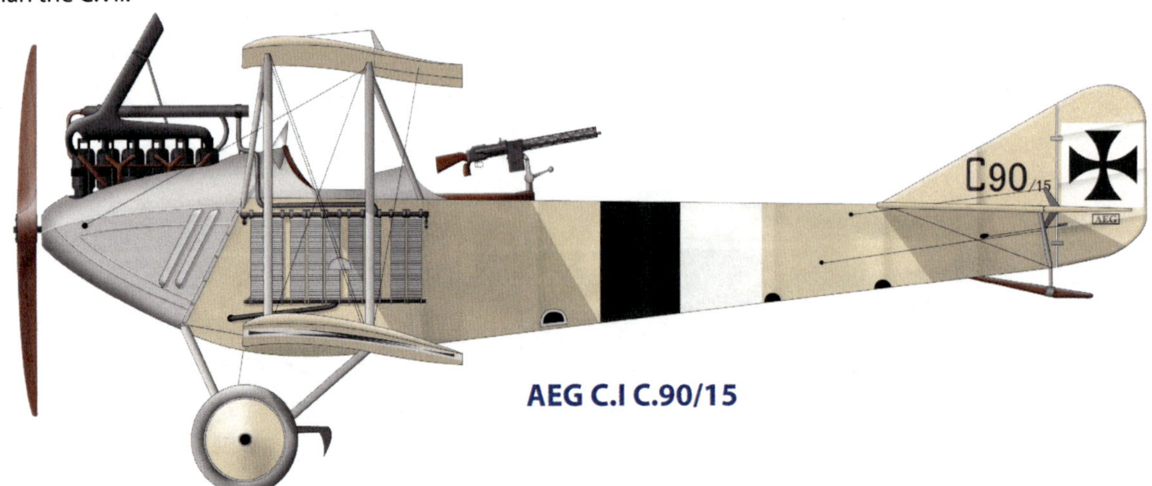

AEG C.I C.90/15

Above: The third AEG C.VIII prototype was built as a triplane. The climb rate was good but like most triplanes it was slower than the biplane from which it was derived, and it did not handle as well. (The Peter M. Bowers Collection/The Museum of Flight)

AEG C.IV C.4871/17 of *FAA* 303 based in Palestine

Ago C.I

Above: The Ago C.I prototype photographed at Johannisthal when rolled out in January/February 1915. The tail had no fixed fins; these were added to production aircraft. The prototypes and first production series had side radiators. Power for the prototype was a 150 hp Benz Bz.III. (Peter M. Grosz Collection/STDB)

The AGO copies of the Otto pusher design had limited performance and failed to excite the interest of the Prussian *Fliegertruppe* who regarded these primitive aircraft as outmoded and were not impressed by the design or workmanship. In fact, the *Fliegertruppe* expressly prohibited the demonstration of new AGO aircraft because they considered it a waste of effort. In addition to all these negatives, the Prussian authorities strongly preferred the tractor aircraft configuration to the pusher configuration.

AGO therefore had strong headwinds to overcome to sell aircraft to the military and had to focus on selling aircraft to the Navy. AGO had built some early seaplanes for the Navy that were overweight, fragile, and not successful. Now the Navy excluded AGO from further consideration, although the Navy did buy the few AGO landplanes available when war broke out and they were desperate for trainers.

Ago C.I Aircraft Production

Order	Serials	Qty	Notes
Feb. 1915	C.1–8/15	8	Side radiators
Feb. 22, 1915	S 158–162	5	For the Navy
Apr. 1915	C.94–105/15	12	Airfoil radiator
1915	C.191–192/15	2	Airfoil radiator
May 1915	C.349–373/15	25	Airfoil radiator
Oct. 1915	C.1903–1914/15	12	Airfoil radiator, two-wheel undercarriage.

Above: Ago C.I C.1/15 after the tailplane was modified with fixed fins; it was then assigned to *Feld-Flieger Abteilung* 9b.

Above: Ago C.I 371/15 of the penultimate production batch.

Above & Right: Ago C.I aircraft of *FFA* 9b modified to carry a long focal length camera. *Lt.* Hans Kuchler was the observer and *Vzfw*. Dorner wss the pilot. (Peter M. Grosz Collection/STDB)

Facing Page: Carl Hailer installing a 120 cm focal length camera in a modified Ago C.I. (Peter M. Grosz Collection/ STDB)

Above: Ago C.I 96/15 was powered by a 160 hp Mercedes D.III. The black and white markings were those of *Armee-Abteilung* Gaede for the purpose of air-to-air identification to avoid attack by other German aircraft. These markings were seen as vertical black/white bands on other aircraft types. (Peter M. Grosz Collection/STDB)

Ago C-Type Specifications				
	Ago C.I	**Ago C.Iw**	**Ago C.II**	**Ago C.IIw**
Engine	150 hp Benz Bz.III 160 hp Mercedes D.III 160 hp Maybach Mb.III	150 hp Benz Bz.III	160 hp Mercedes D.III	220 hp Benz Bz.IV
Span, Upper	15.1 m	15.1	14.50 m (or 18.30 m)	18.3 m
Span, Lower	14.34 m	14.34 m	—	—
Chord, Upper	1.70 m	1.70 m	—	—
Chord, Lower	1.70 m	1.70 m	—	—
Gap	1.74 m	1.74 m	—	—
Wing Area	47.6 m²	47.6 m²	—	—
Length	9.30 m	9.30 m	9.84 m	11.24 m
Height	3.10 m	—	—	—
Empty Weight	960 kg	845 kg	1360 kg	—
Loaded Weight	1,495 kg	1,405 kg	1,946 kg	—
Maximum Speed	130 km/h	130 km/h	137 km/h	—
Climb to 1,000 m	9.5 min.	—	—	—
Climb to 2,000 m	24 min.	—	—	—
Climb to 3,000 m	45 min.	—	—	—

Ago C-Type Specifications				
	Ago C.III	**Ago C.IV**	**Ago C.V**	**Ago C.VI**
Engine	160 hp Mercedes D.III	200 hp Benz Bz.IV	160 hp Mercedes D.III	200 hp Benz Bz.IV
Span, Upper	11.0 m	11.90 m	—	—
Span, Lower	—	—	—	—
Chord, Upper	—	—	—	—
Chord, Lower	—	—	—	—
Gap	—	—	—	—
Wing Area	—	37.5 m²	—	—
Length	7.0 m	8.25 m	—	—
Height	—	3.50 m	—	—
Empty Weight	—	900 kg	—	—
Loaded Weight	—	1,350 kg	—	—
Maximum Speed	—	190 km/h	—	—
Climb to 1,000 m	—	—	—	—
Climb to 2,000 m	—	—	—	—
Climb to 3,000 m	—	22 min.	—	—
C.IV ceiling 5,500 m.				

Ago C.VII engine 200 hp Benz Bz.IV; no further information.
Ago C.VIII engine 260 hp Mercedes D.IVa; no further information.
Ago C.IX engine 260 hp Mercedes D.IVa; no further information.

Ago C.I 96/15 of *Feld-Fleiger Abteilung* 9b. The pilot was *Lt.* Auer and the observer was *Oblt.* Paulin. Their names were painted on the rudder. The nose has been modified to carry a long-focal length camera.

Ago C.I 97/15 of *Feld-Fleiger Abteilung* 9b. The pilot was *Vzfw.* Bruno Chorbacher and the observer was *Lt.d.R.* Ludwig Schmidt. They were based at *Flugplatz Colmar*.

Above: The take-off of an Ago C.I of *Feld-Flieger Abteilung* 8b. The aircraft has a two-wheel undercarriage. (Bruno Schmäling)

Ago C.II

Above: Overhead view of the Ago C.II. The three-bay, long-span wings are clearly shown. (Peter M. Grosz Collection/STDB)

The Ago C.II was a slightly refined, somewhat more powerful development of the Ago C.I with enlarged wing. The 150 hp Benz Bz.III was replaced by a 160 hp Mercedes D.III and the 2-bay wing of the C.I was replaced by a 3-bay wing. An airfoil radiator was fitted as used by later-production Ago C.I aircraft and the airframe was slightly refined.

The modest power increase and airframe refinement provided a small speed increase.

The future did not belong to pusher aircraft and few examples of the C.II were built and used. According to the *Frontbestand* a maximum of two were at the front at the end of 1915 and a few were used by the Navy.

Above: Ago C.II LF.67. was a Naval aircraft.

Above: S.66 was an Ago C.II assigned to the Navy. The text on the nose appears to read "*Schneewittchen*" or 'Snow White.' This name was applied to a number of early German two-seaters that were finished in a pale color. (Peter M. Grosz Collection/STDB)

Above: Another view of Ago C.II S.66 assigned to the Navy was made into a postcard. (Michael Düsing)

Ago C.III

Above: The Ago C.III was a single-bay biplane powered by a 160 hp Mercedes D.III. (Peter M. Grosz Collection/STDB)

Ago's last pusher design was the C.III. It was a small, single-bay biplane powered by a 160 hp Mercedes D.III engine.

The more compact aircraft did not offer enough performance improvement over the C.I and C.II, and only one prototype of the Ago C.III was built; there were no production aircraft. Unfortunately, little information on the type is available. However, the *Frontbestand* chart for C-types indicated the Ago C.III was at the front in the February 1916 inventory, so the prototype was apparently evaluated at the front.

By early 1916 the time of the pusher configuration was over and all additional Ago designs were tractor configuration.

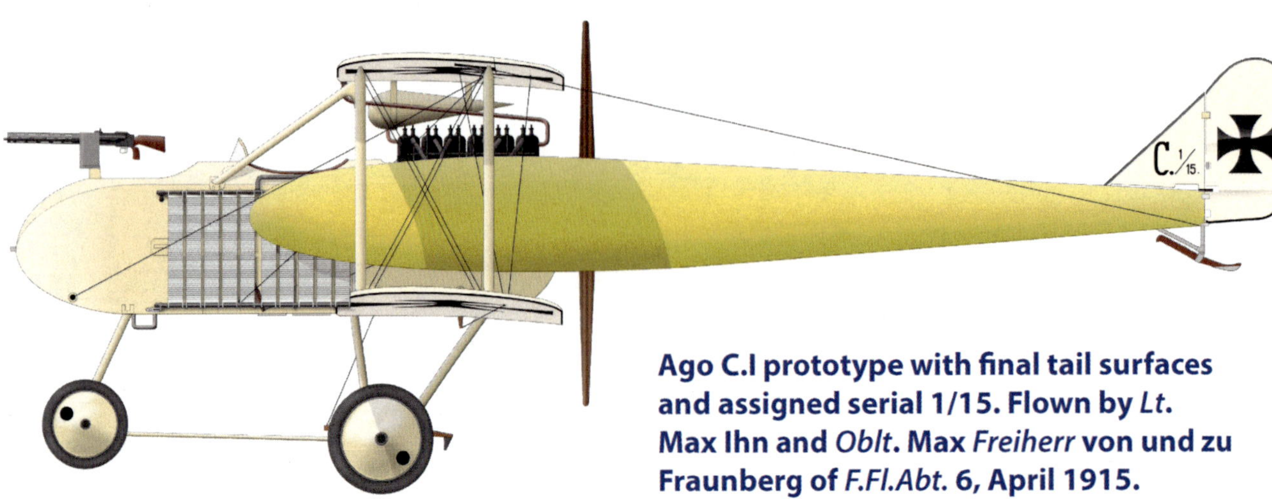

Ago C.I prototype with final tail surfaces and assigned serial 1/15. Flown by *Lt.* Max Ihn and *Oblt.* Max *Freiherr* von und zu Fraunberg of *F.Fl.Abt.* 6, April 1915.

Above & Below: The Ago C.III was a single-bay biplane powered by a 160 hp Mercedes D.III. Only a single prototype was built. (above Peter M. Grosz Collection/STDB)

Ago C.IV

Above: Ago C.IV 2747/16 is an early machine in the initial production configuration from the first production batch. Camouflaged in a two color scheme, there is no fixed fin, ailerons are fitted on the upper wing only, and the broad interplane struts are painted the underside colors. The forward inboard interplane struts are omitted to enable the observer a larger field of fire forward and the rectangular wire in front of the observer's cockpit is to prevent the observer firing into the propeller arc. The fixed machine gun for the pilot is mounted over the engine. (Bruno Schmäling)

Ago had exhausted the potential of the pusher configuration with its C.I, C.II, and C.III designs. To achieve better performance and to create a design that could be sold to the Prussian-dominated *Fliegertruppe*, Ago turned to the tractor configuration.

The result was the Ago C.IV, an innovative design of conventional construction that offered much better speed and climb than the pushers that preceded it.

The Ago C.IV was most easily distinguished from other two-seaters in having a sharply-tapered wing planform. Most WWI aircraft had untapered wings of rectangular planform for ease of manufacturer. However, the induced drag (drag induced by lift) depends on the wing planform, and theoretical calculations indicate that an elliptical wing planform minimizes induced drag. (The Spitfire was an example of an actual design using this principle.) An elliptical planform wing is difficult to manufacture. However, it can be approximated by the tapered wing planform, which is a compromise. The tapered wing has less induced drag than a rectangular wing but is harder to build. It has more drag than an elliptical wing but is easier to build.

The resulting C.IV with 220 hp Benz Bz.IV engine was faster than most two-seat reconnaissance airplanes. It was carefully designed to have clean lines and be relatively streamlined for the time. The combination of good power, uncluttered lines, and tapered wing planform gave it good speed and climb for its time. It was much faster than the C.I that Ago had been building, and had the tractor configuration preferred by the *Fleigertruppe*. Ago appeared to have found the formula, and the C.IV was ordered in quantity starting in June 1916.

Actual design of the Ago C.IV had started in late 1915. Another innovative feature of the C.IV was the lack of a forward, inboard bracing strut to provide a better field of fire forward between the wings for the observer. Ago was given Patent No. 299,149 for this innovation on Oct. 7, 1915, as they were designing the C.IV.

The Ago C.IV prototype flew in early 1916 and immediately demonstrated good performance. In trials flown on May 21 the prototype reached 4,000 m altitude in 32 minutes, fast for the time. This encouraged *Idflieg* to order 24 aircraft in June 1916. Contemporary *Idflieg* files referred

Above & Below: The prototype Ago C.IV photographed at the Ago factory at Johannisthal in spring 1916. It was finished in an overall light color, not camouflaged. The nose contours and lack of propeller spinner also differentiated it from production aircraft. Other than that was similar to the initial production configuration except for being unarmed. There is no fixed fin, ailerons are fitted on the upper wing only. The forward inboard interplane struts normally seen on two-bay aircraft were omitted to enable the observer a larger field of fire forward; however, the rectangular wire in front of the observer's cockpit to prevent the observer firing into the propeller arc that was fitted to production aircraft was not fitted to the prototype. Although of conventional wood, wire, and fabric construction, the Ago C.IV had a number of innovative features and had a significantly better performance the Ago's earlier pusher designs. (Peter M. Grosz Collection/STDB)

Above & Below: The first production version of the Ago C.IV during type-testing at Adlershof in June 1916. Like the prototype, there is no fixed fin and ailerons are fitted on the upper wing only. Furthermore, it has a propeller spinner and revised nose contours. The forward inboard interplane struts normally seen on two-bay aircraft were omitted to enable the observer a larger field of fire forward. The broad interplane struts were painted in the undersurface color, likely light blue. However, unlike the light-doped prototype, this aircraft had the production two-tone camouflage scheme. The scheme was not very distinct and was green and red/brown. The Ago C.IV had a number of innovative features, some of which had to be modified after operational service, and had a significantly better performance the Ago's earlier pusher designs. (Peter M. Grosz Collection/STDB)

Ago C.IV Aircraft Production			
Order	**Serials**	**Qty**	**Notes**
June 1916	2740–2763/16	24	Ago
Sept. 1916	3714–3723/16	10	Rathgeber (not completed; reduced to spares)
Oct. 1916	5500–5749/16	250	Schütte-Lanz*
Nov. 1916	3600–3623/16	24	Ago
Nov. 1916	8950–8999/16	50	Ago
Feb. 1917	1521–1570/17	50	Ago
May 1917	5575–5674/17	100	Ago. Few if any delivered. Order transferred to LVG C.II(Ago) trainers in Sept. 1917.

* Order reduced to 150 in July 1917 & cancelled in Oct. 1917 after 66 aircraft were delivered.

to the Ago C.IV as *the so-called light C-type*. It was thus a precursor of the later CL-class.

Type-testing was completed in mid-June. The testing revealed the need to wing strengthening, and after two reinforcements of the metal wing fittings the required strength was achieved.

An order for 10 aircraft was given to the Munich-based Waggonfabrik Josef Rathgeber AG to bring that company into much-needed aircraft production. Initially the order was for the Rathgeber C.I but was later re-designated the Ago C.IV(Rat) when *Idflieg* rationalized aircraft designations. However, these aircraft were never delivered and the parts were reduced to spares.

Spare production capacity was available at Schütte-Lanz, best known for its airship manufacture as a competitor to Zeppelin. By this time airships were losing favor and the superior Zeppelins could fill requirements. None of Schütte-Lanz's own aircraft designs were selected for production and *Idflieg* ordered 250 Ago C.IV(Schül) aircraft from them to utilize their spare production capacity. The size of the order indicated *Idflieg's* interest in the C.IV.

In December 1916 the Ago C.IV arrived at the front. The pilots, who had been used to the stable and reliable Ago C.I, were anticipating the C.IV to combine these qualities with significantly improved performance. However, they were soon disappointed. As soon as the C.IV went into service, a rash of fuselage failures occurred during landing. However, this problem was quickly corrected in January and deliveries continued.

As the crews logged increasing flight time on the C.IV, fatalities due to side-slipping increased alarmingly. The Ago C.IV was grounded at the end of April pending investigation and solution of the flight-control problem.

The original C.IV prototype and production aircraft had ailerons on the upper wing only and did not have a fixed fin. As a result of the accidents the flight controls of the C.IV were extensively redesigned to improve its flight characteristics. Ailerons were fitted to all wings and a fixed vertical fin was added to the tail. In addition, most all the aircraft delivered by Ago after serial 8950/16 and all built by Schütte-Lanz had an additional bracing strut between the outer wing struts. *Idflieg* also tested Ago C.IV 1521/17 fitted with conventional two-bay bracing. The first Schütte-Lanz built C.IV was type-tested May 7–8, 1917, during which the metal wing fittings failed again and had to be reinforced. Cleared for combat, few Schütte-Lanz built C.IVs reached the front.

The tapered wing planform gave good performance but contributed to the flight-control problem and the technology of the time was not adequate to 'tune' the wing's aerodynamics for good stability. Wings want to stall first at their tips, and this tendency is worsened by tapering the planform. This caused the loss of roll control at the critical moment of a stall. With most rectangular planform wings, the designers 'washed out' the angle of incidence at the wingtips to ensure the wing stalled first at the wing root, countering its tendency to stall first at the tips where the ailerons were located, ensuring the pilot retained aileron control during stall recovery.

After the modified C.IV was returned to the front in the May–June 1917 timeframe, the C.IV was moderately successful but was still unpopular with aircrews due to its tarnished reputation. Continuing pilots' resistance and flight-control problems caused *Idflieg* to ground the C.IV again on September 7, 1917 and cancel production.

The Ago C.IV was the last Ago design to be placed in production.

Above & Below: Ago C.IV(Schül) 5513/16. The aircraft is in final production configuration including fixed fin, ailerons on all wings connected by an actuating strut, and a diagonal bracing strut added to the outer interplane strut assembly. The fuselage text reads: *Zulässige Belastung bei vollem Tank 520 kg.* (Permissable load at full tank 520 kg) *Leergewicht 1020 kg.* (empty weight 1020 kg). This may indicate training use. (Peter M. Grosz Collection/STDB)

Ago C.V

Above: The Ago C.V was a single-bay biplane using N-struts. The observers' cockpit in this version is raised above the level of the pilot's cockpit to provide the gunner a field of fire forward. (Peter M. Grosz Collection/STDB)

The next Ago design was the compact C.V two-seater powered by the 160 hp Mercedes D.III. The C.V looked significantly different from the earlier C.IV and was more compact because it was designed as a CL-type. The single-bay design used N-struts and the wing planform was nearly rectangular with ailerons on all four wings. The angle of incidence of the wingtips was washed out so the ailerons would stall last, preserving the pilot's aileron control after a stall. The C.V represented a major effort to improve flight characteristics after the serious problems with the earlier Ago C.IV.

Not only was the Ago C.V designed for improved flight characteristics, but it was well-streamlined.

Despite its C-type designation, the compact, single-bay Ago C.V was designed for the CL role but no production was undertaken.

Ago C.IV with stylized 'XS' personal marking. Possibly in late production configuration; can't tell for certain from reference photo.

Above: The Ago C.V with modified rear fuselage contours. (Peter M. Grosz Collection/STDB)
Below: Front view of the Ago C.V. (Peter M. Grosz Collection/STDB)

Ago C.VI

Above & Below: The Ago C.VI prototype was a conventional two-bay two-seater. (Peter M. Grosz Collection/STDB)

The Ago C.VI was a larger, more conventional two-bay, two-seater design than the compact C.V. It was powered by a 200 hp Benz Bz.IV The two-bay design reverted to plain I-struts and again the wing planform was nearly rectangular with ailerons on all four wings.

The Ago C.VI was more conventional that earlier Ago C-types, but the C.V remained a prototype with no production being undertaken.

Ago C.VII

Above: The Ago C.VII in its initial configuration was built in 1916. (Peter M. Grosz Collection/STDB)

The Ago C.VII was basically a late-production Ago C.IV with conventional two-bay wing of rectangular planform. Like the Ago C.IV, the C.VII was powered by a 200 hp Benz Bz.IV.

The C.VII prototype was modified with a vertical fin and rudder that resembled those of the C.V and C.VI. Postwar the prototype was sold to Estonia along with a couple of Ago C.IV aircraft.

Above: The Ago C.VII with revised fin and rudder in Estonia with postwar Estonian markings.

Ago C.VIII

Above: The Ago C.VIII prototype was a conventional two-bay two-seater built in 1917.

Below: A later prototype of the Ago C.VIII with refined fuselage and exhaust..

The Ago C.VIII was another conventional two-bay two-seater with rectangular wing planform. Reportedly the engine was a 260 hp Mercedes D.IVa.The wood fuselage was sheathed with plywood.

The photos show modifications to the C.VIII for improved flight characteristics and performance. At top the fin and rudder are more pointed; the other photos show a more rounded fin and rudder profile.

The photo at the bottom on the facing page shows a very different wing bracing design with three bays of slanted struts. The change lowered the upper wing closer to the top of the fuselage.

Above: The later Ago C.VIII prototype had an airfoil radiator mounted in the center section. Ailerons were mounted on all four wings and the wings had rectangular planform. This aircraft was the Ago C.VIII/I or CL.8/I C.1114/18. (Peter M. Grosz Collection/STDB)

Above: Modified Ago C.VIII prototype in front of the Ago factory at Johannisthal. The wood wheels reflected the rubber shortage. The wing cellule has been dramatically redesigned with three bays of slanted struts. (Peter M. Grosz Collection/STDB)

Ago C.IX

Above: The Ago C.IX prototype was a conventional two-bay two-seater. It looked like the Ago C.VIII with airfoil radiator replaced by a leading edge radiator.

The final Ago C-type design was the C.IX. The C.IX appears to have the same airframe and engine (260 hp Mercedes D.IVa) as the C.VIII but with a leading edge radiator replacing the airfoil radiator installed in the C.VIII.

Like the earlier C.VIII, the C.IX was a conventional two-bay design with wood fuselage wrapped with plywood. The wings had conventional struts and rectangular planform. Ailerons were fitted to all four wings.

The C.IX remained a prototype and no production was undertaken.

Left & Below: The Ago C.IX prototype was a conventional two-bay two-seater.

Ago C.IV(Schul) 5513/16 in late production configuration.

Albatros C.I of *Vzfw*. Emil Thuy, *FAA* 53. Thuy went on to become a successful fighter pilot, commanding *Jasta* 28 and ultimately *Jagdgruppe Nr.*7. He used his initial to identify this C.I and his later aircraft and survived the war with 35 victories and was awarded the *Pour le Mérite*.

Albatros C.I 110/15 of an unknown unit on the Russian Front. It carried an over-wing Madsen machine gun in addition to its normal observer's Parabellum and also carried Mauser carbine in event of a forced landing.

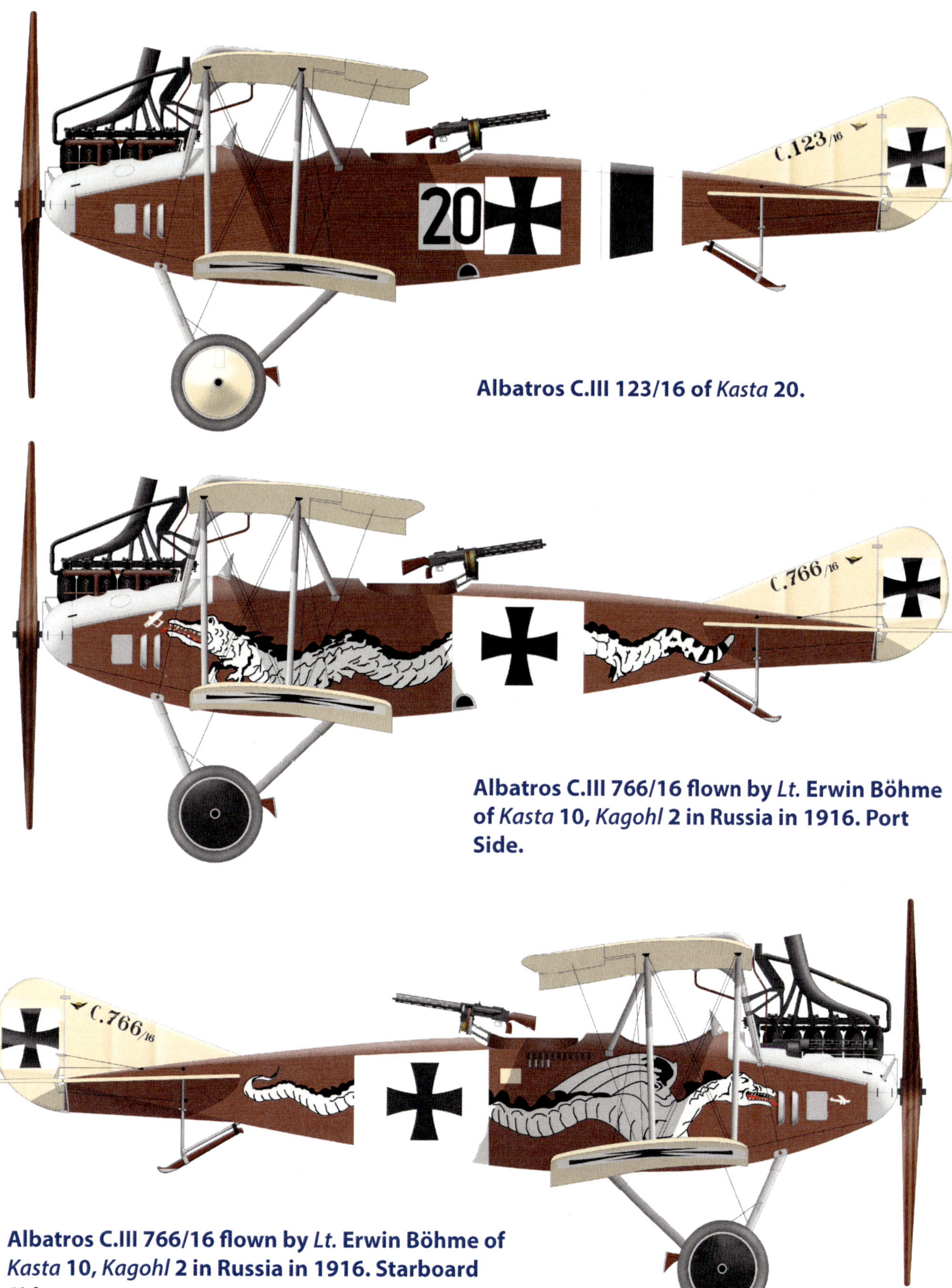

Albatros C.III 123/16 of *Kasta* 20.

Albatros C.III 766/16 flown by *Lt.* Erwin Böhme of *Kasta* 10, *Kagohl* 2 in Russia in 1916. Port Side.

Albatros C.III 766/16 flown by *Lt.* Erwin Böhme of *Kasta* 10, *Kagohl* 2 in Russia in 1916. Starboard Side.

Albatros C-Types

Above: This early production Albatros C.I was powered by a 160 hp Mercedes D.III cooled by *Hazet* side radiators manufactured by Haegele & Zweigle. Like many Albatros C.I aircraft it carries no serial number.

The first generation of Albatros C-types was actually the third generation of Albatros reconnaissance airplanes, the first being the primitive A-type monoplanes and the second being the B-type biplanes. The Albatros C.I was easily developed from the B.II by fitting a more powerful engine, either a 150 hp Benz Bz.III or 160 hp Mercedes D.III, and a flexible gun for the observer, who was moved to the rear cockpit to provide him with a better field of fire.

The C.I set the pattern for the following first-generation Albatros C-types. The Albatros C.III, which was designed as an improved C.I, was developed from the unarmed B.III by combining the best features of the C.I and B.III. Like the development of the C.I from the B.II, a more powerful engine was installed together with a flexible machine gun for the observer, who was now located in the rear cockpit. However, most C.III aircraft, which had few parts directly interchangeable with the B.III, also carried a fixed gun for the pilot.

The C.I and C.III were the most widely-used of the first generation Albatros C-types. Other first-generation Albatros C-types included the C.II that was an experimental pusher, and the C.IV aircraft that were essentially C.Is except the gunner was in the front seat. Finally, the C.VI was a C.III airframe powered by a 180 hp Argus As.III that was used in moderate numbers.

The second generation of Albatros C-types included the C.V, C.VII, and C.VIIIN. The C.V was a specialized high-speed reconnaissance plane built around the rare 220 hp Mercedes D.IV eight cylinder engine. It used a more streamlined fuselage than the first Albatros C-type generation and was faster than Allied fighters when it first arrived in combat. The C.V was quickly followed by the C.VII, a very similar aircraft powered by the 200 hp Benz Bz.IV and intended for general reconnaissance duties. The C.VIIIN was a special long-span, three-bay design intended for night bomber duties. It reverted to the 160 hp Mercedes D.III to conserve resources but remained a prototype.

The third generation of Albatros C-types included the C.X and C.XII, both powered by a 260 hp Mercedes D.IVa engine; no C.XI was built. The C.X and C.XII were both large, powerful aircraft intended for long-range reconnaissance. However, the competing Rumpler C.IV series offered much better performance and the C.X and C.XII were relegated to general reconnaissance duties and training.

The fourth and final generation of Albatros C-types included the C.XIV and C.XV. These were smaller, more compact and maneuverable aircraft powered by the 220 hp Benz Bz.IVa engine. The C.XV was a refined version of the C.XIV and went into production late in the war.

The C.IX and C.XIII were specialized designs intended for the light C-type (that is, CL-type) role and combined the technology of second generation C-types with the 160 hp Mercedes and configuration of the Albatros fighters. Neither was chosen for production, the competing Halberstadt and Hannover types having better handling and maneuverability combined with generally superior performance.

All Albatros C-types were similar in construction, using conventional wood, wire, and fabric flying surfaces, a semi-monocoque plywood fuselage, and a water-cooled inline engine, all but the C.V having six cylinders. They were robust, well-streamlined aircraft with generally good flying qualities,

Albatros C.I Specifications				
	Albatros C.I	**Albatros C.Ia(Bay)**	**Albatros C.Ib**	**Albatros C.If**
Engine	150 hp Benz Bz.III 160 hp Mercedes D.III	180 hp Argus As.III	160 hp Mercedes D.III	160 hp Mercedes D.III
Span, Upper	13.08 m	12.90 m	12.896 m	12.996 m
Span, Lower	11.28 m	11.12 m	—	—
Chord, Upper	1.80 m	1.80 m	1.80 m	1.80 m
Chord, Lower	1.80 m	1.80 m	1.80 m	1.80 m
Gap	1.66 m	1.65	1.66 m	1.66 m
Wing Area	—	40.48 m^2	—	38.04 m^2
Wing Dihedral	2°	1.75° (upper & lower)	—	—
Wing Sweepback	0.5° (upper)	1.70° (upper & lower)	—	—
Length	7.80 m	7.85 m	7.81 m	7.85 m
Height	2.96 m	3.014 m	3.04 m	3.23 m
Track	—	2.25 m	2.21 m	2.207 m
Empty Weight	840 kg	910 kg	875 kg	839 kg
Loaded Weight	1,350 kg	1,415 kg	1,380 kg	1,154 kg
Maximum Speed	132 km/h	140 km/h	140 km/h	140 km/h
Climb to 1,000m	9.8 minutes	13 minutes	9 minutes	9 minutes
Climb to 2,000m	25 minutes	31 minutes	—	22 minutes
Climb to 3,000m	58.5 minutes	—	43 minutes	43 minutes

and most had competitive performance when they reached the front.

Perhaps surprisingly, many Albatros C-types were built simultaneously for slightly different roles. A key strategy of *Idflieg*, driven by its commander, *Oblt.* Wilhelm Siegert, was emphasis on different tasks or roles by different aircraft, with the aircraft optimized for that role. One result of this policy was the wide variety of aircraft typically seen in most German two-seater units. Different aircraft types were provided to the units to fly in the different roles they performed. In other words, the different aircraft types specialized in different roles because they were specifically designed for those roles.

Not only were units equipped with different aircraft types, another result was that different aircraft types, built for different roles and powered by different engines, were built in parallel. Albatros and the other manufacturers that built Albatros types under license were simultaneously building the Albatros C.III (Mercedes D.III engine), Albatros C.V (Mercedes D.IV engine), Albatros C.VI (Argus As.III engine), Albatros C.VII (Benz Bz.IV engine), and Albatros C.X (Mercedes D.IVa engine).

Development of a new type did not necessarily mean that an existing type would go out of production. Old types were taken out of production when replaced by a newer type that was superior in the same role. Furthermore, production of different types depended on engine availability. This was especially evident in the case of the Albatros C.V that was powered by the rare Mercedes D.IV. Limited numbers of the C.V were ordered due to the small number of engines for it. Availability of the new Mercedes D.IVa enabled production of the C.X and C.XII that were intended for the same role as that in which the C.V excelled, long-range photo reconnaissance. Despite offering disappointing performance, these aircraft were built in larger numbers than the successful C.V because the engine they used was available in much greater quantity.

Finally, airframes were much easier to build than engines, which required precision machinery, skilled labor, and high-quality materials in short supply. When an airframe was obsolete, typically the engine would be recycled into another, superior aircraft designed to use it.

Albatros C-Type Specifications

	Albatros C.II	Albatros C.III	Albatros C.IV	Albatros C.V
Engine	150 hp Benz Bz.III	150 hp Benz Bz.III 160 hp Mercedes D.III	160 hp Mercedes D.III	220 hp Mercedes D.IV
Span, Upper	—	11.00 m (short span) 11.70 m (long span)	—	12.78 m
Span, Lower	—	11.04 m (short span) 11.14 m (long span)	—	12.44 m
Chord, Upper	—	1.80 m	—	1.80 m
Chord, Lower	—	1.70 m	—	1.80 m
Gap	—	1.65 m	—	1.83 m
Wing Area	—	34.37 m^2 (short span) 37.10 m^2 (long span)	—	43.4 m^2
Wing Dihedral	—	1.75° (upper & lower)	—	2° (upper & lower)
Wing Sweepback	—	1.70° (upper & lower)	—	—
Length	—	7.95 m	—	8.95 m
Height	—	3.07 m	—	4.5 m
Empty Weight	—	771 kg	—	1,069 kg
Loaded Weight	—	1,271 kg	—	1,585 kg
Maximum Speed	—	135–150 km/h	—	170 km/h
Climb to 1,000m	—	5 min best, 8 min avg	—	4.5 minutes
Climb to 2,000m	—	12.5 min best, 22 min avg	—	9.5 minutes
Climb to 3,000m	—	25 min best, 45 min avg	—	16 minutes
Climb to 4,000m	—	—	—	25 minutes

The refined Albatros C.III was intended as an improved C.I and used the same 150 hp Benz Bz.III or 160 hp Mercedes D.III engines. The C.III was more streamlined and had better performance coupled with the addition of a fixed pilot's gun (except for early production aircraft), all of which made it a more effective combat aircraft than the C.I.

Albatros C-Type Specifications					
	Albatros C.VI	Albatros C.VII	Albatros C.VIIIN	Albatros C.IX	Albatros C.X
Engine	180 hp Argus As.III	200 hp Benz Bz.IV	160 hp Mercedes D.III	160 hp Mercedes D.III	260 hp Mercedes D.IVa
Span, Upper	11.7 m	12.78 m	16.74 m	10.4 m	14.36 m
Span, Lower	—	12.40 m	—	—	14.00 m
Chord, Upper	—	1.80 m	—	—	1.8 m
Chord, Lower	—	1.70 m	—	—	1.6 m
Gap	—	1.83 m	—	—	1.86 m
Wing Area	—	43.4 m^2	—	—	42.7 m^2
Wing Dihedral	—	2° (upper & lower)	—	—	2° (upper & lower)
Length	7.9 m	8.71 m	7.34 m	8.22 m	9.15 m
Height	3.2 m	3.60 m	—	2.735 m	3.40 m
Empty Weight	830 kg	1,030 kg	—	790 kg	1,088–1,115 kg
Loaded Weight	1,343 kg	1,546 kg	—	1,150 kg	1,668–1,695 kg
Maximum Speed	145 km/h	135 km/h	135 km/h	155 km/h	175 km/h
Climb to 1,000m	—	5.5 minutes	5 minutes	5 minutes	3 minutes
Climb to 2,000m	—	13 minutes	—	—	6.5 minutes
Climb to 3,000m	35 minutes	21 minutes	—	—	11 minutes
Climb to 4,000m	—	34 minutes	—	30 minutes	21 minutes
Climb to 5,000m					49 minutes
Duration	4.5 hours	—	—	2.5 hours	3½ hours
Note: C.VII track 1.95 m					

Below: The streamlined Albatros C.V/16 was the first of the second generation Albatros C-types. The eight-cylinder Mercedes was completely enclosed by the cowling, the key C.V recognition feature. The tail surfaces and fuselage strongly resembled contemporary Albatros fighters. Bruno Loerzer (in sweater) stands in front of this C.V/16(OAW) of *Kampfstaffel* Metz; *Hptm.* Schmickaly is in the front cockpit.

Albatros C-Type Specifications					
	Albatros C.XII	**Albatros C.XII(OAW)**	**Albatros C.XIII**	**Albatros C.XIV**	**Albatros C.XV**
Engine	260 hp Mercedes D.IVa	260 hp Mercedes D.IVa	160 hp Mercedes D.III	220 hp Benz Bz.IVa	220 hp Benz Bz.IVa
Span, Upper	14.37 m	14.24 m	10.00 m	10.4 m	11.8 m
Span, Lower	13.01 m	12.92 m	—	—	—
Chord, Upper	1.8 m	1.8 m	—	—	—
Chord, Lower	1.6 m	1.6 m	—	—	—
Gap	1.73 m (inboard) 1.80 m (outboard)	1.73 m (inboard) 1.80 m (outboard)	—	—	—
Wing Area	42.7 m²	42.7 m²	—	—	—
Wing Dihedral	3° (upper), 2.5° (lower)	3° (upper), 2.5° (lower)	—	—	—
Length	8.85 m	8.85 m	7.8 m	6.9 m	7.47 m
Height	3.40 m	3.40 m	2.71 m	—	3.33 m
Empty Weight	1,020–1,115 kg	1,040–1,115 kg	700 kg	950 kg	859 kg
Loaded Weight	1,638–1,733 kg	1,658–1,733 kg	1,060 kg	1,385 kg	1,320 kg
Maximum Speed	175 km/h	175–180 km/h	165 km/h	—	165 km/h
Climb to 1,000m	5 minutes	5 minutes	4 minutes	—	3.4 minutes
Climb to 2,000m	11 minutes	11 minutes	—	—	—
Climb to 3,000m	19 minutes	19 minutes	—	—	—
Climb to 4,000m	30 minutes	30 minutes	—	—	—
Climb to 5,000m	45 minutes	45 minutes	47 minutes	—	—
Climb to 6,000m	—	—	—	—	47 minutes
Duration	3¼ hours	—	2.5 hours	3.5 hours	3 hours
Note:					

Below: The Albatros C.VII closely resembled the Albatros C.V that was designed at nearly the same time. However, the 200 hp Benz Bz.IV protruded above the fuselage whereas the C.V's Mercedes engine was fully enclosed.

Albatros C.I Production Orders

Serials	Type & Notes
C.15/15 to 26/15	Alb C.I
C.44/15 to 67/15	Alb C.I
C.106/15 to 129/15	Alb C.I
C.192/15 to 246/15	Alb C.I
C.249/15 to 273/15	Alb C.I
C.443/15 to 492/15	Alb C.I
C.550/15 to 639/15	Alb C.I
C.640/15 to 663/15	Alb C.I(Rol) = Rol C.I
C.1000/15 to 1024/15	Alb C.I
C.1067/15 to 1114/15	Alb C.I (incomplete?)
C.1450 /15to 1499/15	Alb C.I
C.1529/15 to 1578/15	Alb C.I
C.1800/15 to 1835/15	Alb C.I(Rol) = Rol C.I
C.1986/15 to 2003/15	Alb C.I(Rol) = Rol C.I
C.13375/17 to 13574/17	Alb C.Ib(Mer)
C.15000/17 to 15199/17	Alb C.Ia(Bay)
C.15500/17 to 15599/17	Alb C.Ia(Bay)
C.4900/18 to 4999/18	Alb C.Ib(Mer)
C.5000/18 to 5149/18	Alb C.IbRin)
C.5520/18 to 5669/18	Alb C.Ia(Bay)
C.9200/18 to 9399/18	Alb C.Id(Mer) & C.If(Merc)

Albatros C.III Production Orders

Serials	Type & Notes
C.2060/15 to 2103/15	Alb C.III
C.4000/15 to 4199/15	Alb C.III
C.100/16 to 193/16	Alb C.III
C.650/16 to 715/16	Alb C.III(Bay) = Bay C.I
C.716/16 to 815/16	Alb C.III
C.1375/16 to 1404/16	Alb C.III
C.2124/16 to 2173/16	Alb C.III(Bay) = Bay C.I
C.2274/16 to 2374/16	Alb C.III(OAW)
C.200/17 to 239/17	Alb C.III(Bay) = Bay C.I
C.3428/17 to 3527/17	Alb C.III(Bay) – trainer
C.4000/17 to 4099/17	Alb C.III(Hansa)
C.5000/17 to 5199/17	Alb C.III(LVG)
C.5400/17 to 5474/17	Alb C.III(Li)
C.6200/17 to 6299/17	Alb C.III(Bay)
C.7127/17 to 7426/17	Alb C.III(Bay) – could also be 7125 to 7424
C.12000/17 to 12199/17	Alb C.III(OAW)
C.13375/17 to 13574/17	Alb C.III(DFW)
C.14000/17 to 14099/17	Alb C.III(SSW)
C.15200/17 to 15299/17	Alb C.III(LVG)
C.15600/17 to 15649/17	Alb C.III(Hansa)
C.1115/18 to 1164/18	Alb C.III(Hansa)

Albatros C.IV Production Order

Serials	Type & Notes
C.850/15 to 861/15	Alb C.IV

Right: The nose of a C.X third generation Albatros C-type displays it powerful 260 hp Mercedes D.IVa six-cylinder engine. The C.X greatly resembled the C.VII from which it was derived and the types shared some of their components. The C.X did use a new, larger wing cellule to support the larger, heavier engine. While evolutionary design served Albatros well through the C.VII, something more innovative was needed for the third generation C-types using the 260 hp Mercedes.

Albatros C.I

Above: The Albatros C.I was a transitional design, retaining the basic airframe of the B.II with a more powerful engine and rotating gun turret for the observer, who was moved to the rear cockpit to maximize his field of fire. This C.I has a 160 hp Mercedes D.III engine. (Peter M. Bowers Collection/The Museum of Flight)

As air-to-air combat became more common, it was obvious that reconnaissance aircraft needed effective defensive armament, which meant a flexible machine gun for the observer who needed to be relocated to the rear cockpit for maximum field of fire. In turn, a more powerful engine was needed to maintain performance with the additional weight of the machine gun and ammunition.

Albatros followed this strategy with development of the C.I from the unarmed B.II. The C.I had a flexible gun fitted in the rear cockpit and moved the pilot to the front. The C.I was also fitted with a 150 hp Benz Bz.III or 160 hp Mercedes D.III in place of the 100–120 hp engines used in the B.II. Use of both engine types was necessary to maximize production. Early production C.I aircraft retained the side radiators of the B.II, but later production aircraft received a radiator mounted above the engine.

The Albatros C.I was the first aircraft of this new configuration to reach the front. The Aviatik C.I reached the front simultaneously but retained the gunner in the front cockpit, a configuration that was not nearly as effective and was soon abandoned.

In addition to its flexible gun for self defense, the C.I also had internal bomb racks for four 10 kg bombs. Early production aircraft carried 1.5 mm armor on the fuselage floor and 2.5 mm armor for the pilot and observer per *Idflieg* specification, but this was of limited value and reduced performance so was soon abandoned. Cameras and wireless were also fitted as needed.

The C.I retained the steady reliability and strength of the B.II from which it was derived and also inherited its good flight characteristics. Both the C.I's structure and aerodynamics remained conventional like its B.II predecessor. Overall the C.I was a very utilitarian, practical, evolutionary design, not an innovative breakthrough.

The C.I was built in small numbers under license by Roland in Germany and by the Oesterreichisch-Ungarische Albatros Werke GmbH near Vienna for the Austro-Hungarian *Luftfahrtruppe*. In Austro-Hungarian service the C.I was designated Albatros B.I(Ph) series 23, 24, and 22 in that order.

Like the B.II from which it was derived, the C.I was robust and had good flying qualities, so

Above & Below: Albatros C.Ia(Bay) C.15001/17 photographed at Adlershof during its flight testing in December 1917. Built by BFW (Bayerische Flugzeug Werke) in Munich for training use, the C.Ia(Bay) was powered by a 180 hp Argus As.III. The leading edge radiator is by Windhoff and the wing struts and wheels are made of wood to conserve materials. The flying surfaces were covered with lozenge camouflage fabric. (Peter M. Bowers Collection/The Museum of Flight)

Above: C.I 637/15 gets an assist taking off from a snowy field.

Right: The C.I was primarily designed with straight lines.

when retired from the front survivors were used as advanced trainers. Also like the B.II, additional C.Is were ordered built new as trainers, and these served to the end of the war. C.I trainers were built by BFW as the Albatros C.Ia(Bay); these aircraft used the 180 hp Argus As.III engine. BFW-built C.I trainers had strengthened wing spars and wooden wing struts, the additional strength enabling the drag wire from the nose to be eliminated. Because the C.I was intended as an advanced trainer, the training versions had camera and wireless installations.

During the type test of the C.Ia(Bay) in January 1918 the rear fuselage failed. External stiffeners were applied to existing airframes and stronger longerons were used in subsequent production aircraft.

Mercur Flugzeugbau GmbH received orders for the Albatros C.Ib, a trainer powered by the 160 hp Mercedes D.III, and later for the Albatros C.Id(Mer) and C.If(Mer). The C.Ib had wooden wing struts and some had pneumatic springs by Hofmann in the landing gear legs. During the type test the Mercur-built aircraft also suffered rear fuselage failure and had to be reinforced similar to the C.Ia(Bay). It is thought that the C.Id(Mer) indicated dual control and the C.If(Mer) indicated use of the Hofmann pneumatic springs.

Above: Land-based naval aviators "Franz und Emil" pose in front of their Albatros C.I LF127 (LF for *Landflugzeug*) powered by a 150 hp Benz Bz.III. The Navy had 25–26 Albatros C.I aircraft, of which 10 were powered by the 150 hp Rapp Rp.III. Originally the Navy used an "S" designation for *Schulflugzeug* (school airplane, or trainer), but on 7 March 1915 the "S" designation was changed to mean any naval land airplane. On 18 November 1915 the S-designation was changed to LF.

Above: Mercedes-powered Albatros C.I with side radiators lacked a visible serial number. A practical design, the C.I was a key first-generation armed reconnaissance airplane.

Above & Below: Anonymous Benz-powered Albatros C.I aircraft at the front.

Albatros C.II

Above & Below: The Albatros C.II is shown above after full markings were added and below at the factory at Johannisthal. (Peter M. Bowers Collection/The Museum of Flight)

The Albatros C.II was a definite departure from typical Albatros practice. German manufacturers wisely avoided the pusher configuration with the high-drag *Gitterschwanz* (lattice tail), but the C.II was apparently intended to assess that configuration. The C.II had a 150 hp Benz Bz.III and apparently used wings from the C.I. Only one prototype, C.27/15, was built in early 1916.

Above & Below: The Albatros C.II pusher at the Albatros factory at Johannisthal. Propeller blast has splattered the rudder with dirt. The pusher configuration gave the observer an exceptional field of view and field of fire forward if he were placed in the front cockpit as in most pushers, but pushers were very vulnerable to attack from the rear where the observer had no field of fire. (Peter M. Bowers Collection/The Museum of Flight)

Albatros C.III

Above: Albatros C.155/16 with 160 hp Mercedes D.III engine combined features from the successful C.I and B.III. This early aircraft has no fixed gun for the pilot. (Peter M. Bowers Collection/The Museum of Flight)

The C.III was developed as an improved C.I and followed it into production. The C.III was a combination of the features of the C.I and B.III. It used the same engines as the C.I, either the 150 hp Benz Bz.III or 160 hp Mercedes D.III, but was lighter, more compact and more streamlined, and finally abandoned the side radiators for a radiator in front of the upper wing. Of course, construction remained the typical Albatros semi-monocoque plywood fuselage and wood, wire, and fabric wings. Compared to the C.I, the C.III had improved maneuverability and performance, especially climb rate.

In addition to its improved performance and maneuverability, the C.III also had a fixed, synchronized machine gun for the pilot, a feature the C.I lacked. The additional armament was in response to more intense Allied fighter attacks and became standard for German C-types for the rest of the war. From then on German two-seaters were generally difficult opponents for Allied fighters.

The C.III saw widespread, successful service in combat and, like the B.II and C.I before it, remained in production for the rest of the war as a trainer due to its reliability and good handling qualities. Its advantages were soon recognized and extensive license production was undertaken.

A number of C.III reconnaissance airplanes were supplied to Turkey, receiving Turkish serial numbers AK8 to AK41.

Interestingly, for an unknown reason the C.III was built with two different wing spans, 11.00 meters and 11.70 meters.

Above: Albatros C.III C.169/16 was an early production aircraft. The serial was painted on the fin parallel to the fin's leading edge, and the wings and tail surfaces were in clear-doped linen. The fuselage was varnished plywood with the national insignia painted over a white background.

Above & Top: Albatros C.III with Benz engine photographed with a measurement stick for scale. This early-production aircraft has no fixed gun for the pilot.

Right: Albatros C.III C.173/16 with Mercedes engine at *Flieger Abteilung* 2.

Above: Albatros C.III with Mercedes D.III engine and no fixed gun; C.III aircraft with Benz engines are in the background.

Below & Below Right: Closeup of an Albatros C.III with Mercedes D.III engine and Fokker or Albatros interrupter system.

Above: Albatros C.III C.2072/15 was from the first production batch and has no fixed gun for the pilot. This Benz-powered aircraft wears a light overall finish that enables the national insignia to be applied directly without a white background.

Above: Benz-powered Albatros C.III 4047/15 with fuselage in natural finish and over-wing gravity tank.

Above: Mercedes-powered Albatros C.III AK62 in Turkish markings furnished with an oriental rug.

Below: Mercedes-powered Albatros C.III flown by Neisen carries a propeller-driven generator on the forward, starboard landing gear strut to power a wireless set for artillery spotting. Unusually, no fixed gun was installed for the pilot.

Albatros C.IV

Above: Albatros C.IV C.850/15 was used as a test aircraft for the Albatros G.III wing cellule. Although the test cellule was smaller than the G.III, the thick airfoil and distinctive struts of the G.III are clearly seen; the rest of the C.IV was apparently the same as the well-known C.III. However, the pilot's cockpit was in the rear and the observer's cockpit was in front – and interestingly enough, rails for mounting flexible machine guns are on each side of the observer's cockpit!

A small batch of 12 Albatros C.IV aircraft, serials C.850–861/15, were ordered. These aircraft were essentially C.III aircraft with the exception that their observers were in front and pilots in back. This was ordered by *Idflieg* as a test of the benefit of the observer in front; *Idflieg* ordered LVG C.III C.862/15–873/15 with a similar gun arrangement.

By the time the C.IV batch was built the superiority of the gunner in the rear cockpit was proven and the C.IV was likely not used in combat on the Western Front. C.850/15 was used to test the Albatros G.III wing cellule. C.851/15 and C.860/15 were delivered to German flying schools as trainers, and C.853–859/15 (with exception of C.857/15) were delivered to the Turkish air service. At least one of these, C.853/15, was an unarmed mail plane.

Above: Three rare aircraft are captured in this photo. A Pfalz E.VI in the foreground is testing experimental camouflage and a Pfalz E.V is in the background right of center. However, directly behind the E.VI is Albatros C.IV C.850/16 as can be determined by its distinctive wing and interplane struts.

Above: Unarmed Albatros C.IV C.853/15 in use as a mail plane in Mesopotamia with Turkish markings and normal C.III-type wings. Aircraft C.853–859/15 (except for C.857/15) were delivered to the Turkish flying corps. At least two aircraft, C.851/15 and C.860/15, were delivered to German flying schools for use as trainers.

Above & Below: Two images of Albatros C.IV C.850/15 in front of the Albatros factory at Johannisthal. No serial is visible and no gun mounting rails are fitted around the forward observer's cockpit, but an interesting fairing is seen below the observer's cockpit, perhaps for a camera? (Peter M. Bowers Collection/The Museum of Flight)

Above: Albatros C.IV C.850/15 at Johannisthal with gun mounting rails fitted around the sides of the forward observer's cockpit. (Peter M. Grosz Collection/SDTB)

Below: Albatros C.III AK60 in Turkish markings with German pilot Emil Meinecke and Observer *Oberleutnant* Ott of the Ottoman (Turkish) *Fliegerabteilung* 6. The number AK60 is seen on the top of the white border of the fuselage square. Available Turkish records say that the Turkish Albatros C.III aircraft were all numbered AK8 to AK41, but from photographs it is apparent that Turkey got more Albatros C.III aircraft than documented.

Albatros (OAW) C.I

Above: Albatros (OAW) C.I C 9/15 in front of the OAW factory. This early aircraft, the second built, was fitted with side radiators. Based on serial numbers, seven of these aircraft were built. (Peter M. Bowers Collection/The Museum of Flight)

The OAW facility, co-located with the Albatros Militär Fliegerschule GmbH, was originally intended solely for producing designs of the main factory. Regardless of its inception as solely a production facility, OAW did build two different C-type prototypes as a further outlet for their creativity.

Seven examples of the first type, initially known as the OAW C.I and later often referred to as the Albatros (OAW) C.I, were built with serials C.8/15 – C.14/15. It had typical Albatros construction.

The OAW C.I was a large, 3-bay biplane powered by a 150 hp Benz Bz.III. It appears to have used the basic wing from the Albatros B.I with enlarged wing-root cuts for greater field of view downward and featured a large triangular fin and rudder similar to the standard Albatros C.I. Appearing early in 1915, it was an early, fairly unsophisticated biplane. However, it was a practical design that evolved in production despite the small number that were built. The above photo of C9/15, the second aircraft built, shows its primitive side radiators, while C13/15, the sixth aircraft built, has a more advanced, leading edge radiator.

Despite the small number built, there were several variations in the exhaust manifolds fitted. Some exhausted downward and there were at least two manifold configurations that exhausted upward.

The crew members were located close together for mutual cooperation during flight, and the cockpit was especially large and roomy. A large windscreen was fitted for the pilot on the first aircraft built but the others do not appear to have had this. However, there was a large combing in front of the pilot's cockpit that provided limited protection from the wind blast. In fact, the combing was so high pilots had to stretch to see over or around it when looking forward. The observer had a flexible machine gun; however, there was no fixed gun for the pilot.

Available photographs show the OAW C.I apparently in operational use and also in training service, with one photo of the latter taken after April 1918 based on the partially visible national insignia.

Above: Albatros (OAW) C.I C.8/15 was the first aircraft of its type built and carries a flexible gun for the observer. It appears to be in operational service. The large windscreen for the pilot is notable. (Peter M. Grosz Collection/SDTB)

Above: Albatros (OAW) C.I C.13/15 in front of the OAW factory. Built in 1915, this three-bay reconnaissance prototype was powered by a 150 hp Benz Bz.III; this aircraft has a radiator under the upper wing. Although the OAW factory was intended for production only, not design, it is obvious senior people there wanted another outlet for their creativity.

Albatros (OAW) C.II

Above: The OAW C.II shared the basic configuration of the earlier OAW C.I although it was more powerful yet compact. Built in 1916, this two-bay reconnaissance prototype was powered by a 220 hp Mercedes D.IV eight-cylinder engine. The large, four-blade propeller and completely enclosed engine are prominent recognition features. The fuselage cross has the octagonal white background often used by OAW. (Peter M. Bowers Collection/The Museum of Flight)

The OAW facility followed their C.I design with a more refined, powerful design. The OAW C.II was more compact, having a two-bay design instead of the three-bay C.I. Equally significant was the much more powerful engine in the C.II, which used the rare 220 hp Mercedes D.IV eight-cylinder engine in place of the 150 hp Benz used in the C.I.

Like the earlier OAW C.I, the C.II was a practical design that evolved despite the few built. The first C.II prototype had unbalanced control surfaces. However, it was a large aircraft and a subsequent prototype had aerodynamically balanced rudder and elevators to reduce the pilot's control forces to make the aircraft more agile and less tiring to fly. Also like the C.I, the crew members in the C.II were located close together for mutual cooperation during flight and the pilot enjoyed a large, spacious cockpit.

Right: Albatros (OAW) C.II warming up with no one in the cockpit. (Peter M. Grosz Collection/SDTB)

Above: This Albatros (OAW) C.II features revised tail surfaces with horn-balanced rudder and elevators to reduce control forces. The crew had a spacious cockpit and the observer was located close to the pilot to facilitate communication in flight. The pilar mount for the observer's flexible machine gun is visible. A few OAW C.II prototypes were built but the exact number is not known and it was not produced in quantity.

An Albatros (OAW) C.II warming up. The synchronized machine gun for the pilot is clearly visible. (Peter M. Grosz Collection/SDTB)

Albatros C.V

Above: The streamlined Albatros C.V was the first of the second-generation Albatros C-type designs to reach the front. It was also the only C-type without a six-cylinder engine; the C.V used the 220 hp Mercedes D.IV, a straight-eight cylinder engine created by adding two more cylinders to the 160 hp Mercedes D.III. The propeller reduction gear raised the thrust line of the propeller, enabling the engine to be fully cowled. The reduction gear also reduced propeller speed, enabling a large, slow-turning propeller to be fitted for maximum efficiency and greater speed. This aircraft was one of the three C.V prototypes from the batch C3593–3595/15 and was photographed in front of the Albatros factory in March/April 1916. The external, over-wing gravity tank was not fitted to production aircraft.

Availability of new, more powerful engines resulted in the second generation of Albatros C-types. Albatros designers continued to use the same semi-monocoque fuselage construction and wood, wire, and fabric wings while emphasizing improved streamlining. Two new types were developed almost in parallel that used some of the same components, such as wings and tail surfaces.

First to reach the front was the Albatros C.V using the 220 hp Mercedes D.IV eight-cylinder engine. It was closely followed by the Albatros C.VII using the 200 hp Benz Bz.IV engine. The different engines gave these similar aircraft different performance characteristics and consequently they were used for different combat roles.

In January 1915 *Idflieg* gave development contracts for engines of 200 hp or more to Benz, Daimler (manufacturers of Mercedes engines), and Körting. Mercedes began work on two different engines to meet this requirement. The D.IV engine was an enlargement of the 160 hp, six-cylinder Mercedes D.III by the simple expedient of adding two more cylinders. The resulting engine, rated at 220 hp, was the first of the two to become available. The D.IVa engine was a totally new six-cylinder design that took longer to develop; when available it was rated at 260 hp.

The Albatros C.V was designed around the first of these engines to become available, the 220 hp Mercedes D.IV. This straight-eight engine, the only such engine flown operationally during the war, was viewed as an interim engine and only a limited quantity were ordered pending availability of the more advanced Mercedes D.IVa.

The D.IV had an integral propeller reduction gear that had two notable consequences. First, the thrust line of the propeller was raised above the crankshaft. This enabled the engine to be mounted lower in the airframe and allowed it to be fully cowled, improving streamlining. Second, the gear reduced 1400 RPM engine speed to 900 RPM for the propeller, allowing a larger, slower-turning and more efficient propeller to be used.

The result was that the streamlined, more powerful C.V was significantly faster than preceding Albatros C-types. In fact, when it arrived at the front in July 1916 the C.V was faster than any Allied fighter at the front! This was a remarkable result for an essentially evolutionary design that was a product of a powerful new engine combined with careful development by Albatros. This happy result was also

Above: This experimental C.V was likely one of the three prototypes. The I-struts reduced drag; however, they also obstructed the crew's field of view somewhat and that made them unpopular. Moreover, the stress calculations for the I-struts were still only approximate and the engineers were reluctant to proceed with them. This aircraft also had the ear radiators installed with a louvered panel to reduce drag, a feature that was not used in production. The observer's gun run has not yet been installed and an external gravity tank was fitted above the upper wing.

Above: Albatros C/16 1176/16 was the second production aircraft; it was photographed during a quiet moment in the field at *Flieger Abteilung* 2. The aircraft has the early production series features of squared wingtips on the lower wings, ear radiators, and coolant header tank over the engine. The large (3.4 meter diameter), slow-turning propeller used with the geared Mercedes D.IV was also a distinctive C.V characteristic.

the only time an Albatros two-seater design provided such superior performance.

The C.V's speed and climb made it a natural for long-range photo-reconnaissance missions and this was the role in which it excelled. Being larger and substantially heavier than earlier Albatros C-types, it lacked their maneuverability and was not at its best during general-purpose reconnaissance and artillery spotting, where good maneuverability to evade attacking fighters was important. Instead, the C.V was best at using its speed and altitude capability (missions could be flown at 5,200 meters, impressive for 1916) to avoid interception completely, and this it was able to do for many months in combat.

The first production batches ordered in 1916 totaled 75 C.V aircraft. A final batch of 50 was ordered in January 1917 and these differed from the 1916 model in several respects. The aerodynamic improvements developed for the Albatros C.X and C.XII then in design and test were applied to this batch of C.V aircraft. The lower wingtips were rounded, the ailerons were aerodynamically balanced, and an aerodynamically balanced elevator was fitted. The ear radiators in the 1916 model were replaced by an airfoil-shaped radiator in the upper wing center section. Finally, the lighter control forces due to the aerodynamically-balanced controls enabled a control stick to replace the former control wheel. These changes enhanced maneuverability and performance, which helped the revised C.V/17 remain competitive as new Allied aircraft appeared.

To differentiate between the two models, the earlier C.V model is denoted as the C.V/16 and the later model as C.V/17. Some C.V/16 models were given the later wings during factory repairs.

The remaining C.V aircraft were withdrawn from the front in August 1917, and in December there were still 84 C.Vs of the 125 built in inventory, an impressive survival record. A few C.V aircraft were used as advanced trainers after withdrawal from combat, but the simpler, cheaper, easier to fly B.II, C.I, and C.III were much more popular as trainers.

C.V 1420/17 was used as an airborne testbed for the 2 cm Becker cannon. Firing trials were performed in October and November 1917 with a downward-firing Becker fixed in the rear cockpit. The pilot must have been cramped because he shared the rear cockpit with the cannon, while the observer sat in front and apparently loaded the cannon.

Facing Page, Below: A prototype Albatros C.V in front of the Albatros factory at Johannisthal illustrates the characteristic features. The prototypes had a wingspan of 12.2 meters; production aircraft had a longer, 12.78 meter span. The exhaust stack has a different shape than production aircraft, and there is no fixed gun for the pilot. There is also an external gravity tank over the upper wing that was not present in production aircraft and the observer's gun ring is recessed. (Peter M. Bowers Collection/The Museum of Flight)

Above: Albatros C.V/17 photographed on the snow-covered Johannisthal airfield during the winter of 1916/1917. Elimination of the ear radiators in this batch improved its streamlining. (Peter M. Bowers Collection/The Museum of Flight)

Above: Albatros C.V/17 1371/17 shows its refined airframe with airfoil radiator in place of the ear radiators of the C.V/16. The header tank above the engine on the C.V/16 was no longer needed, so the exhaust and fixed Spandau are the only blemishes on the streamlined engine cowling. The upper surfaces of the wings are painted in the three-color camouflage scheme used by later C.V and C.VII aircraft. The fuselage appears to be stained in a dark color and the arrow insignia of *Flieger Abteilung* 7 is just barely visible against the dark fuselage color. The typical signal flare cartridges are mounted alongside the observer's cockpit. (Peter M. Grosz Collection/STDB)

Above: An Albatros C.V/17 in flight reveals its graceful lines including rounded lower wingtips.

Above: During its takeoff an Albatros C.V/17 of *Flieger Abteilung* 2 displays its elegant streamlining.

Albatros C.VI

Above: The Albatros C.VI was a C.III airframe fitted with a 180 hp Argus As.III engine. The Argus had a bit more power than the 160 hp Mercedes used in the C.III, making the C.VI slightly faster at low altitude. However, the Mercedes was lighter and gave better power at altitude, giving the C.III a better climb rate. The Albatros factory at Johannisthal is the background. (Peter M. Bowers Collection/The Museum of Flight)

The final first-generation Albatros C-type to be produced was the C.VI. The C.VI was simply a C.III airframe fitted with a 180 hp Argus As.III engine. The Argus had slightly more power than the 160 hp Mercedes in the C.III but was heavier and lost power more quickly with altitude than the Mercedes. Consequently, the C.VI was slightly faster than the C.III at low altitude but had a lower climb rate and ceiling and was not quite as nimble.

In any case, engine production generally was the main factor limiting aircraft production, so production of the Argus-powered C.VI was a useful supplement to C.III production.

Although the *Frontbestand* inventory for the C.VI peaked at 111 on the end of February 1917, the only known order information was for a batch of 75 aircraft placed in May 1916, so either at least one production batch is unknown or the *Frontbestand* inventory is in error. Because the *Frontbestand* table was created after the war by one individual from multiple wartime reports, it is known to contain some errors; the number of Albatros C.VI aircraft listed at the front in February 1917 is likely one of them.

Above: Albatros C.VI, 1777/16 at the front. Tactical '1' is painted behind the fuselage insignia. (Peter M. Grosz Collection/SDTB)

Above: Albatros C.VI 1816/16 takes off. Attached to an unknown unit, this aircraft wears dark camouflage with prominent national insignia and a swastika-like marking on its wheel cover. Known C.VI serials are C.1775–1849/16.

Above: This photo clearly shows the nose details of an Albatros C.VI, including its Argus engine and synchronized gun installation. The propeller was made by Axial; the dagger with "Axial" written on it was an early Axial logo. (Peter M. Grosz Collection/SDTB)

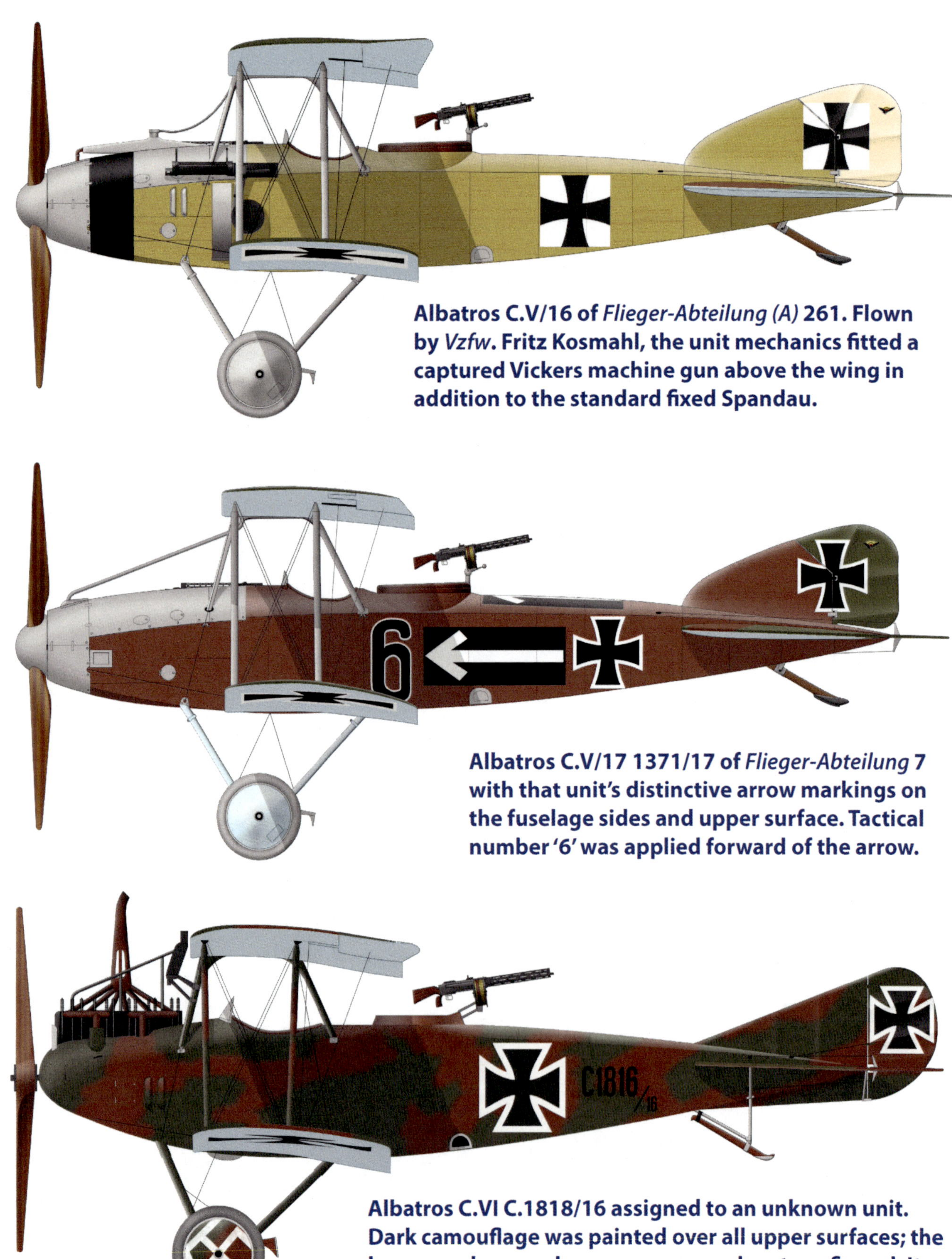

Albatros C.V/16 of *Flieger-Abteilung (A)* 261. Flown by *Vzfw*. Fritz Kosmahl, the unit mechanics fitted a captured Vickers machine gun above the wing in addition to the standard fixed Spandau.

Albatros C.V/17 1371/17 of *Flieger-Abteilung* 7 with that unit's distinctive arrow markings on the fuselage sides and upper surface. Tactical number '6' was applied forward of the arrow.

Albatros C.VI C.1818/16 assigned to an unknown unit. Dark camouflage was painted over all upper surfaces; the brown and green shown are assumed, not confirmed. It is not known if the markings painted on the wheel covers were a personal or unit marking.

Albatros C.VII C.1330/16 of *Flieger-Abteilung* 7.

Albatros C.VII C.2220/16 of *Schusta* 4 in early 1917. *Schusta* 4 was formerly *Kasta* 16 and the aircraft may have been with the earlier unit. A captured Lewis gun was fitted over the wing. The '1' and cross were lighter than the national insignia and are shown here as red.

Albatros C.X(OAW) 9244/16 flown by *Lt*. Hugo Geiger and *Lt*. Theodor Rein of *Flieger Abteilung* 46b in mid 1917. The wings and tailplane are shown with plain under surfaces but they may have been painted blue.

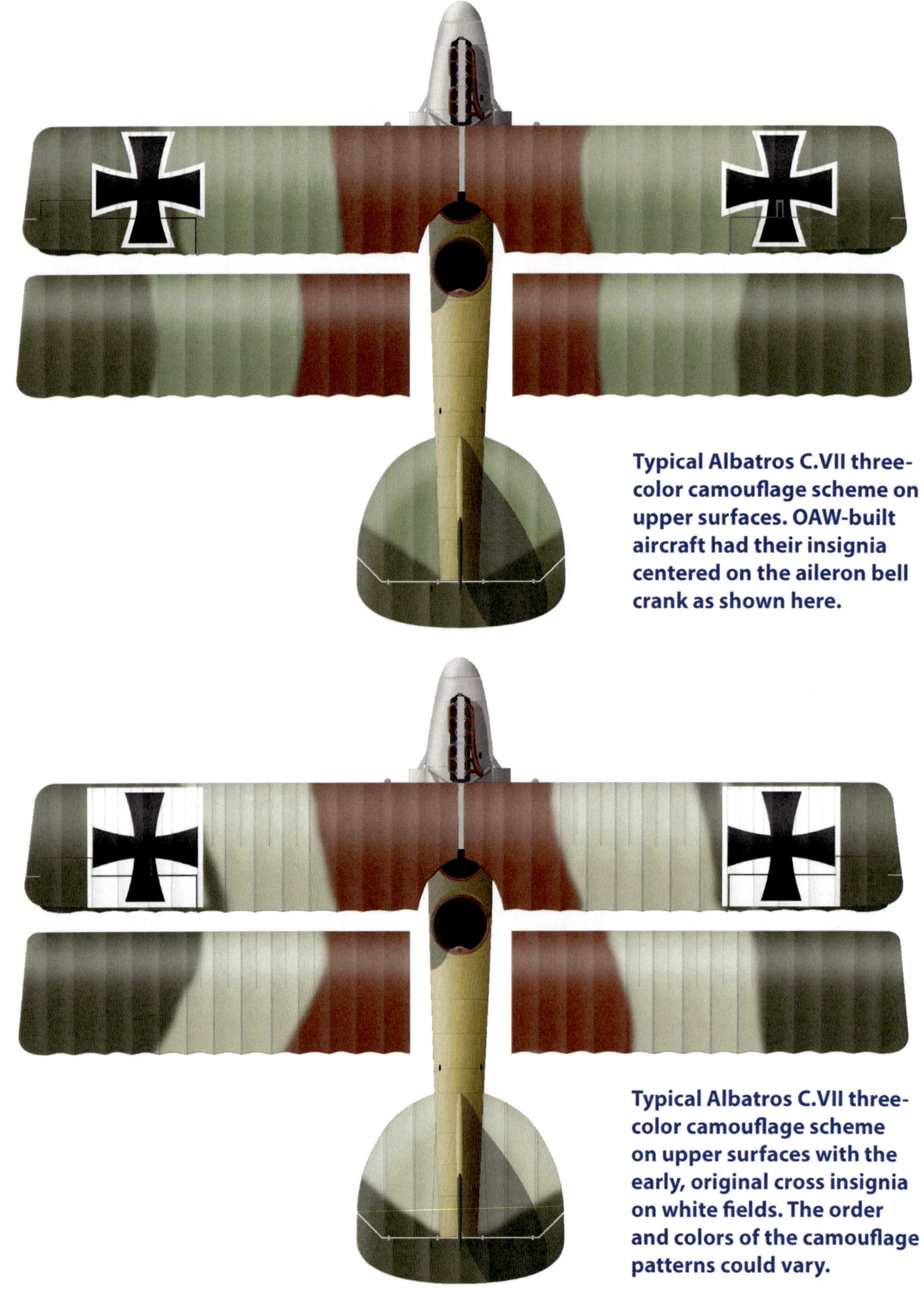

Typical Albatros C.VII three-color camouflage scheme on upper surfaces. OAW-built aircraft had their insignia centered on the aileron bell crank as shown here.

Typical Albatros C.VII three-color camouflage scheme on upper surfaces with the early, original cross insignia on white fields. The order and colors of the camouflage patterns could vary.

Albatros C.VII(Bay) 7713/16 of an unknown unit.

Albatros C.VII(Bay) 7713/16 with camouflage applied to the wings in narrow bands roughly parallel to the wing ribs. BFW-built aircraft had their wing insignia more inboard than Albatros or OAW-built aircraft. There is a possibility the pattern had three colors like those of other manufacturers.

Albatros C.VII

Above: Albatros C.VII 1359/16 at *Armee Flug Park* 3 carries a black and white fuselage band. The upper surfaces of the wings are painted in a three-color scheme and the national insignia have a 50mm wide white surround. Rounded lower wingtips were tested on C.VII prototypes but have not been seen on operational aircraft; C.VII production was complete by the time rounded wingtips were tested.

The Albatros C.VII was designed along with the C.V. Powered by the new 200 hp Benz Bz.IV, a standard production engine of which 7,124 were built during the war, the C.VII was intended for mass production for general-purpose reconnaissance and light bombing duties.

The C.V and C.VII shared their general appearance and many parts. The C.V, C.VII, and later C.X shared their fin, rudder, horizontal tailplane, and elevators. *Idflieg* claimed the C.V and C.VII also shared wing cellules, although the ailerons were different, the C.V/16 having plain ailerons and the C.V/17 and C.VII having balanced ailerons. However, the upper wings of the C.V and C.VII were the same in span, airfoil section, and planform. Similarly, the lower wings of the C.V/16 and C.VII were the same, while the C.V/17 lower wing had a different planform with rounded wingtips.

Like the C.V/16, the C.VII had ear radiators. The key visual difference between the C.V and C.VII was the C.VII had exposed engine cylinder heads while the C.V had its engine fully enclosed. The C.V/17 also featured rounded lower wingtips whereas the lower wings of the C.VII had the same squared-off planform of the C.V/16.

With its lower power engine, less efficient propeller, and greater drag of its exposed engine, the C.VII was not nearly as fast as the speedy C.V, nor could it match the climb rate and ceiling of the C.V. The C.V routinely flew its long-range reconnaissance missions at 5,200 meters; the service ceiling of the C.VII was about 4,000 meters. While the C.V could often soar above and out run Allied fighters during its long-range photographic missions, the C.VII typically had to fight its way to or from its tasks of medium-range reconnaissance, light bombing, and artillery spotting. Not as handy as the lighter C.III, the C.VII had better climb and ceiling that made interception more difficult.

The C.VII appeared over the front in September/October 1916, shortly after the C.V, and by February 1917 the C.VII was the most numerous C-type in combat. However, production of the C.VII was completed by the end of 1916 and by April 1917 it was outnumbered by the superior DFW C.V. Although powered by the same engine as the Albatros C.VII, which essentially disappeared from combat by the end of 1917, the DFW C.V served at the front in large numbers until the war's end.

Above: This aircraft is thought to be the Albatros C.VII prototype. It was photographed at Johannisthal in front of the Albatros company hangars. The exposed engine cylinder head above the cowling is the key identification factor differentiating the C.VII from the C.V that was developed at basically the same time. The C.VII shared its squared-off lower wing with the C.V/16 and its upper wing, with balanced ailerons, with the C.V/17.

Below: Albatros C.VII 3540/16 photographed at *Flieger Abteilung (A)* 224 on 22 May 1917. This aircraft has the square white backgrounds on the national insignia that were obsolete by this time. The balanced ailerons and square-tipped lower wings are clearly shown. The wings wear a three-color camouflage scheme and there is a dark chevron marking with tactical number '3' on the fuselage. It is likely this aircraft was actually assigned to *Schusta* 27.

Above: Albatros C.VII reconnaissance aircraft of an unknown two-seater unit.

Below: Albatros C.VII(Bay) 7713/16 from the last production batch ordered in October 1916 was painted in an overall camouflage finish. It was flown by *Flieger Abteilung* Karlshorst.

The C.VII was a strong, stable, well-made aircraft that had sufficient speed, climb, and maneuverability to hold its own in combat until the new generation of Allied fighters powered by the Hispano-Suiza V-8 became numerous in the summer of 1917.

In addition to production by Albatros, BFW produced the C.VII under license. During type testing of the BFW-built C.VII during 12–25 October 1916 the load tests had to be performed five times, with fixes made between tests, before the aircraft passed. During later load tests of a random production BFW-built aircraft the aircraft again failed, and BFW aircraft were noted for their poor workmanship that compromised their safety.

After the C.VII was retired from combat it continued to serve as an advanced trainer until the end of the war. The C.VII was the last successful Albatros two-seater design to see operational service during the war.

Above: Here is a C.VII with a leading edge radiator in place of the normal ear radiators. Two photos exist of C.VII 1330/16 of *Flieger Abteilung* 7 with this field modification, and this may be another photo of that aircraft; the *Flieger Abteilung* 7 arrow unit marking is just visible next to the observer's cockpit. A captured Lewis is mounted on the upper wing.

Below: This C.VII of *Flieger Abteilung* 7 retains its ear radiators but has had a captured Lewis gun attached to the cabane struts. A close look at the aircraft above reveals similar mounting points on the cabane struts. Was the aircraft above the same aircraft as below after radiator modifications, or was more than one aircraft of the unit fitted with a Lewis?

Albatros C.VIIIN

Above, Below, & Bottom: The Albatros response to the N-type requirement was the C.VIIIN, a typically-streamlined Albatros that looked much like a longer-span, 3-bay Albatros C.XII. Here it is shown carrying the required six 50 kg *PuW* bombs totalling 300 kg. However, with only 160 hp compared to the 260 hp of the C.XII, it was much too under-powered. More power might have given it competitive performance, but Albatros apparently did not explore that.

Albatros submitted the Albatros C.VIIIN for the N-type requirement. Well streamlined with great resemblance to later Albatros two-seaters, the C.VIIIN looked like a longer wing-span, three-bay version of the Albatros C.XII. However, powered by a 160 hp Mercedes D.III engine instead of the 260 hp Mercedes D.IVa used in the C.XII, it was much too under-powered for operations. However, a more powerful engine was apparently not tested, and the AEG N.I with 150 hp Benz Bz.III was built instead.

Albatros C.IX

Above: Colorized view of Richthofen in Albatros C.IX 4508/16 at Hannover, Germany, ca. Sept/Oct, 1917. The Albatros C.IX used the technology of the Albatros fighters and fortunately avoided the single-spar lower wing of the D.III and D.V/Va. It was designed to meet *Idflieg's* specification for a light C-type. Over-shadowed by the competing designs from Halberstadt and Hannover, it did not go into production. Manfred von Richthofen used this example for personal transportation and may be the aviator in the cockpit.

In August 1916 *Idflieg* issued technical specifications for lightened C-type aircraft powered by 160–180 h.p. engines. The light C-type, or CL-type, was intended to be a two-seat escort fighter, short-range reconnaissance aircraft and artillery spotter, and ground-attack aircraft. Two companies, Halberstadt and Hannover, created excellent CL-designs that entered production in mid-1917 and served successfully until the armistice. Albatros also designed a couple of aircraft to this requirement, but these did not enter production.

The Albatros C.IX was built in 1917 and appears to have been designed with the CL-specification in mind. The C.IX was a single-bay biplane that was notably smaller than earlier Albatros two-bay C-types. Unsurprisingly, the C.IX used the same semi-monocoque wood fuselage structure as the successful Albatros fighters and was powered by the same 160 hp Mercedes D.III engine. However, the wing design featured two spars for both upper and lower wings with X-struts in addition to parallel struts for robust interplane bracing. These interplane struts appear heavier than needed. The lower wing was straight with rounded tips, while the upper wing was mounted forward with a large cut-out over the pilot for better visibility and access, and was sharply swept-back to maintain the proper center of lift. Although not fitted, the C.IX was designed to mount the standard German two-seater armament of a fixed Spandau machine gun for the pilot and a flexible Parabellum machine gun for the observer.

The Albatros C.IX did no go into production and only three were built. It is mainly known for its association with the Red Baron, Manfred von Richthofen, who briefly used one as a squadron hack; however, one served with *Flieger Abteilung* 12 in 1918.

Above & Below: In response to *Idflieg's* requirement for a light C-type the Albatros C.IX exhibited excellent workmanship. Design of the C.IX was also more innovative than most Albatros aircraft, although it had a streamlined fuselage characteristic of Albatros aircraft of that time. However, the competing Hannover CL.II was a more compact design that offered better speed and climb and, together with the Halberstadt CL.II, was selected for production in place of the shapely C.IX. (Peter M. Bowers Collection/The Museum of Flight)

Comparison of Albatros C.IX & C.XIII to Hannover CL.II			
	Albatros C.IX	**Albatros C.XIII**	**Hannover CL.II**
Engine:	160 hp Mercedes D.III	160 hp Mercedes D.III	180 hp Argus As.III
Wing Span	10.4 m (34.1 ft.)	10.0 m (32.8 ft.)	11.95 m (39.2 ft.)
Length	8.22 m (27.0ft.)	7.8 m (25.6 ft.)	7.8 m (25.6 ft.)
Empty Weight	790 kg (1,742 lb.)	700 kg (1,543 lb.)	773 kg (1,704 lb.)
Loaded Weight	1,150 kg (2,535 lb.)	1,060 kg (2,337 lb.)	1,133 kg (2,498 lb.)
Maximum Speed	155 km/h (97 mph)	165 km/h (103 mph)	165 km/h (103 mph)
Climb to 1000 m	5 min.	4 min.	4 min.
Climb to 4000 m	30 min.	–	–
Climb to 5000 m	–	47 min.	32 min.
Armament:	1 flexible machine gun & 1 fixed machine gun	1 flexible machine gun & 1 fixed machine gun	1 flexible machine gun & 1 fixed machine gun

Above: The excellent Hannover CL.II was placed in production. It had better speed and climb than the Albatros C.IX and was also noted for its maneuverability, something the C.IX was unlikely to have matched. The Albatros C.XIII fared better in comparison but had a single-spar lower wing that caused structural problems in Albatros fighters.

Albatros C.IX 4508/16 serving as a squadron 'hack' in *JG* I and used by Manfred von Richthofen in September/October 1917.

Albatros C.IX prototype mid 1917.

Above: Albatros C.IX photographed at *Flieger Abteilung* 12 in February 1918. *Idflieg* normally ordered three prototypes of experimental aircraft, one for flight testing, one for structural testing, and a spare. The structural test airframe was normally tested to destruction, leaving two flyable airframes. Manfred von Richthofen used one as a transport, and here is the other. Being continually short of aircraft the German Navy often assigned prototypes to operational service; that was much less common for the Army. Regardless, here is a C.IX prototype at an operational unit – and it night camouflage. Was it used for light bombing? Night fighting? Night reconnaissance? Or all those roles? (Peter M. Grosz Collection/SDTB)

Albatros C.IX photographed at *Flieger Abteilung* 12 in February 1918 with lozenge night camouflage on the fuselage and hexagonal night camouflage on the wings and horizontal tail.

Albatros C.X

Above: The Albatros C.X prototype retained the ear radiators of the earlier C.VII combined with the new, more powerful 260 hp Mercedes D.IVa engine. Like the preceding C.VII, prototypes had ailerons on the upper wings only.

The next generation of Albatros C-types, the C.X and C.XII, were characterized by their use of the newly available 260 hp Mercedes D.IVa six-cylinder engine. The C.X greatly resembled the earlier C.VII; the two types shared their configuration and typical Albatros structure. The prototype C.X aircraft even used ear radiators like the C.VII, but production C.X aircraft used an airfoil radiator in the upper center section.

When the Mercedes D.IVa engine became available in the second half of 2016, *Idflieg* published technical requirements for a new long-range reconnaissance plane using the new engine. The Albatros C.X and Rumpler C.IV proposals were selected for prototyping, and construction of the C.X began in August or September 1916.

The C.X used a larger wing than the C.V and C.VII, and for the first time Albatros used a new box spar design driven by the shortage of aircraft-quality lumber. Other than that, the C.X was basically an enlarged C.VII.

Both the C.X and competing Rumpler C.IV had flown by 2 October 1916. Load tests on C.X airframe work number 2910 were performed at Adlershof during 10–14 October. The airframe passed four of the five load cases. A strengthened wing as tested on 21 October and again failed. Once the ribs behind the rear spar were strengthened the wing passed a third attempt on 21 November.

Production C.X aircraft differed from the prototypes in having airfoil radiators and ailerons on all four wings. The lower wingtips were also rounded, perhaps as a result of observing the rounded lower wingtips on the Rumpler. However, the C.X adhered to the typical Albatros design and construction, and many C.X parts such as the tailplane, elevator, rudder, and parts of the landing gear and controls were common to the C.V and C.VII. The return line for the coolant from the wing radiator was led through the forward port center-section strut for improved streamlining, a detail copied from the LVG C.IV. *Idflieg* had prohibited this by the time the C.X was type-tested due to cooling problems caused by reduced water flow; regardless, many C.X aircraft had this feature, likely due to use of parts built before the prohibition. At least one C.X, C.6831/16, was fitted with an eight-cylinder Mercedes D.IV engine and lasted into 1918.

During prototype evaluation it soon became clear that the Rumpler C.IV was far superior to the Albatros C.X. Despite that, 400 C.X aircraft were ordered from Albatros, Roland, BFW, and Linke-Hofmann, with OAW building most, and perhaps all, of the Albatro-built aircraft. At first C.X aircraft built by BFW and Roland were not rated highly by *Idflieg*, but the problems were resolved; Linke-Hofmann fared better.

Above: Albatros C.X(Li) 8306/16 was one of the first C.X aircraft built under license by Linke-Hofmann. Intended for operational training, the aircraft built by Linke-Hofmann had full operational equipment.

Below: Albatros C.X(Bay) 7760/16 at Adlershof. Although wearing the unit insignia of *Flieger Abteilung (A)* 209, it was flown to Adlershof by a Bavarian air service staff officer on official business.

Initially the C.X was relegated to training use, hardly an effective use of the powerful new Mercedes engine, but aircraft reached squadron service in March/April 1917 in parallel with the Rumpler C.IV. The C.X aircraft that saw combat were from the final OAW production batch, although it is not known why earlier production aircraft were not used at the front. The Rumpler C.IV quickly established itself as the premiere German long-range reconnaissance aircraft. In contrast, the C.X had inferior climb and ceiling compared to the Rumpler and was typically slower. Its disappointing performance caused the C.X to be relegated to general-purpose duties with regular two-seater units, where it showed no real advantage over the 200 hp DFW C.V despite supposedly being faster. The C.X was difficult to fly and disappointing in performance, and was soon withdrawn from the front and allocated to training units.

Above & Below: Albatros C.X(OAW) 9244/16 that served with *Flieger Abteilung* 46b at Marrimbois Ferme. *Lt.* Hugo Geiger was the pilot and the lightning bolt was his personal insignia; the observer was *Lt.* Theodor Rein. The hatch for installation of a long focal-length camera is behind the observer's cockpit in the photo below; in the photo above the camera is installed and can be seen protruding above the top of the fuselage. (Below photo courtesy Bruno Schmäling)

Above: Crewmen *Lt.* Krohl and *Lt.* Kerp pose with their Albatros C.X(OAW) 9252/16 that served with *Flieger Abteilung* 46b at Marrimbois Ferme. The hatch for installation of a long focal-length camera is behind the observer's cockpit. Bullet holes in the tail and rear fuselage have circular patches painted with 'smiley faces' except the 'smiles' were straight, not curved.

Below: Albatros C.X(Bay) of *Flieger Abteilung* 2. The radiator cooling water flows through the forward left cabane strut, a design intended to minimize drag. However, this led to leakage problems and was soon banned by *Idflieg*.

Above: Unidentified Albatros C.X(OAW) with unknown crew shows the streamlined shape of this mediocre aircraft.

Above: Albatros C.X(Rol) 6831/16 was photographed on 16 March 1918 with the new national insignia on fuselage and tail but with the original insignia on the wings. This aircraft had been converted to use the 220 hp Mercedes D.IV eight-cylinder engine. In the process it lost its spinner and gained a new, streamlined metal engine cowling giving it a distinctively different look from the front. The rubber shortage has affected the aircraft by eliminating regular tires. The 260 hp Mercedes D.IVa engine used by the C.X and C.XII was in demand for other operational types and some of the C.X and C.XII aircraft used for training had their D.IVa engines replaced by the D.IV straight-eight. The last new aircraft using the D.IV engine, the rare Lübeck-Travemünde F.2, was delivered in February 1918 and other types using this rare engine had essentially disappeared from operational service, making limited numbers of the D.IV engine available.

Albatros C.XII

Above: The Albatros C.XII was the final evolution of the traditional Albatros C-type designs. Powered by a 260 hp Mercedes D.IVa engine like its C.X ancestor, it was undoubtedly one of the most elegant designs of the time. If only its performance had lived up to its appearance! (Colorization by Jim Miller.)

The Albatros C.XI was apparently not built, the next Albatros two-seater being the C.XII. The C.XII was a refined derivative of the C.X whose development began in December 1916. The C.XII retained the engine, wings, and structural technology of the C.X combined with a more refined fuselage of more streamlined shape. A look at the Albatros C.VIIIN shows a fuselage and tail very similar to the C.XII, and the C.XII fuselage apparently was derived from the C.VIIIN, with longer nose to accommodate the larger engine and a longer tail to balance it. This was similar to the process used to develop the D.V fighter from the D.III and was similarly ineffective. The empty weight of the C.XII was reduced from the 1,088 kg of the C.X to 1,020 kg, a useful but not dramatic change that made little impact on performance.

Like the C.X, the C.XII was equipped with the standard two-seater armament of a fixed gun for the pilot and flexible gun for the observer. Cameras and wireless could be installed, and the C.XII could also carry light bombs. The streamlined fuselage had less keel surface so a larger vertical tail supplemented with a fin under the tail were used for stability. The rounded fuselage required longer landing gear legs and the track was increased to 2.20 meters from the 1.95 meters of the C.X for more stability during ground handling. The C.XII cabane strut arrangement was significantly different from that of the C.X and, along with the presence of the under-fuselage fin, is a useful recognition feature distinguishing the two types.

One of the C.XII prototypes, C.9314/16, was retro-fitted with an eight-cylinder Mercedes D.IV engine, perhaps to evaluate using these engines in C.XII training machines.

The first batch of 300 C.XII aircraft was ordered from Albatros in January, 1917, before the type test had been performed, which began on 1 March. Albatros started delivering C.XII aircraft in May and OAW and BFW-built C.XII deliveries began in June 1917.

In service the elegant C.XII proved little faster than the C.X. Climb, ceiling, maneuverability, and handling were likely somewhat better due to the C.XII's reduced weight, although this is not reflected in the specifications. This was true of the Johannisthal-built aircraft but OAW-built aircraft proved heavier than those built at Johannisthal, and this

Above: Prototype C.XII 9313/16 displays its camouflage scheme; the contrast between the dark green and brown finish is more visible in this view. The stub wing built into the lower fuselage for attachment of the lower wing is easily seen. Developed at the same time as the D.V fighter, the same process of evolutionary refinement was used for both designs. Fortunately, both top and bottom wings had two spars.

Below: The chocks are removed and ground crew hold back prototype C.XII 9313/16 as it readies for takeoff with pilot *Lt.* Lindemann and observer *Lt.* Stier. *"Versuchs-Flugzeug"* (test aircraft) is stenciled below the observer's cockpit.

Above: A C.XII prototype in a hangar at Alt-Auz *Flugplatz*, Latvia. Windhoffer, who managed the airborne cameras for reconnaissance there, is in the light uniform.

Above: On 10 June 1917 C.XII 9313/16 crashed after take off, killing *Lt.* Lindemann and badly injuring *Lt.* Stier after Lindemann lost control during a steep turn at low altitude.

extra weight reduced their climb and ceiling to that of the C.X. Despite having the same engine as the Rumpler C.IV and a nearly identical empty weight, the Rumpler was more maneuverable with better handling and slightly faster with far superior climb and ceiling.

Albatros had slowly evolved their designs from the C.V through C.X to C.XII. In contrast, Rumpler had taken a bold new approach with their 200 hp C.III and endured major development issues with that aircraft. This paid of when availability of the 260 hp Mercedes enabled the Rumpler C.IV based on the C.III airframe developed with such difficulty. Now Rumpler had a much more efficient wing cellule than Albatros that accounted for the performance difference.

In service the C.XII suffered serious radiator leakage problems. The radiator housing was a stressed part of the cabane structure and flight loads were transmitted to the radiator housing, causing the soldered seams to leak. By itself this restricted flight duration and required constant repairs.

Despite its disappointing performance, the C.XII was built in significant numbers and saw service on both the Western and Eastern Fronts. Combat experience soon revealed the C.XII was too vulnerable for long-range reconnaissance missions on the Western Front. First it was relegated to general-purpose, shorter-range reconnaissance duties, then it was removed from active service there, although it continued in service on the less demanding Eastern Front into 1918. Operational service also revealed the streamlined fuselage of the C.XII was not as robust as earlier Albatros C-types and many C.XII aircraft were destroyed in landing accidents, a heavy landing often resulting in the fuselage being broken in two – or even into three pieces.

Once removed from the front the C.XII was used for training duties, where together with the C.X it continued to demonstrate more difficult handling qualities than earlier Albatros C-types.

Above: C.XII(OAW) 1200/17, the first OAW-built C.XII, photographed at Johannisthal in April/May 1917.

Above: Like the Albatros C.X, the Albatros C.XII was tested with an eight-cylinder Mercedes D.IV, almost certainly to evaluate its suitability to power C.XII trainers. Here the third C.XII prototype, 9314/16, is shown fitted with a Mercedes D.IV in mid-late 1918. Like the earlier C.X conversion, the spinner was eliminated and the engine cowling was replaced with a new, streamlined metal cowl. The rubber shortage has affected this aircraft by eliminating rubber tires; that and the number '6' on the fuselage indicate training use. The Mercedes D.IV and D.IVa were nearly the same size and weight and of similar power, making the conversion straight-forward. Use of the obsolescent D.IV in some C.X and C.XII aircraft used for training freed up supply of the 260 hp Mercedes D.IVa engine urgently needed for priority operational types.

Above: Albatros C.XII 1057/17 was flown by pilot *Lt.* Hugo Geiger and observer *Lt.* Theodor Rein of *Flieger Abteilung* 46b. The red lightning bolt was Geiger's personal marking. The covered hatch behind the observer is for installation of a long focal-length camera. *FA* 46b flew from Marinbois Ferme in the late spring and summer of 1917 while equipped with the Albatros C.X and C.XII. Geiger's C.XII 1057/17 was lost on 23 July 1917; engine failure caused an emergency landing that broke the fuselage in two, a common problem with the C.XII.

Above & Below: Albatros C.XII 1109/17 was stationed in Russia, probably with *Flieger Abteilung (A)* 218. The above view shows it carrying two 50 kg PuW bombs under the observer's cockpit, a practice creating an aft center of gravity and reduced stability. The dark patch just in front of the fin is a reinforcement patch to strengthen the rear fuselage, a known weak spot in the C.XII design. The front quarter view below shows the under-fuselage bomb racks for 12.5 kg PuW bombs, a panel in front of the cylinder heads, perhaps to improve airflow, and the rounded compass housing under the lower right wing.

Albatros C.XIII

Above: Colorized photo (by Jim Miller) of the Albatros C.XIII prototype. The streamlined C.XIII was obviously derived from the Albatros D.V fighter. It was much lighter and more compact than the earlier C.IX and was a more convincing design. Although its performance matched the competing Hannover CL.II, it may not have matched the Hannover's excellent maneuverability. Moreover, limited by its single-spar lower wing that had already suffered structural failures in the Albatros D.V, the C.XIII clearly lacked the Hannover's robustness, an essential quality for a successful combat aircraft.

After the failure of the Albators C.IX to reach production for the light C-type role, Albatros created another design to the same *Idflieg* specification. The resulting C.XIII, had a single-bay configuration like the C.IX and, obviously based on the Albatros D.V fighter, was intended for the CL-role. A compact, streamlined aircraft, it was powered by the same 160 hp Mercedes D.III used in the D.V fighter and earlier C.IX prototype. The cockpits were very close for crew coordination. Unfortunately, the C.XIII retained the single-spar lower wing already proved structurally unsatisfactory in the Albatros D.V. Armament was the standard German two-seater fit of a fixed Spandau machine gun for the pilot and a flexible Parabellum machine gun for the observer.

Normally *Idflieg* would order three examples of a new prototype. However, only a single C.XIII was apparently built.

Albatros' second attempt at a light C-type, the C.XIII, was more convincing than its C.IX. It was more refined, with no protrusions, and was much lighter than the C.IX or Hannover. It looked aggressive, and, except for high-altitude climb rate, its performance matched the Hannover. However, to achieve its light weight the C.XIII retained the single-spar lower wing of the D.V, so undoubtedly inherited its structural weakness, whereas the Hannover was a very robust design. Furthermore, the C.XIII may not have matched the Hannover's excellent maneuverability. With a more robust two-spar lower wing like that of the Pfalz D.III, the C.XIII might have been competitive.

Albatros C.XIV

Above: The Albatros C.XIV prototype represented a new direction in Albatros C-type design.

After the relative failure of the large, elegant C.XII, Albatros reconsidered their design direction for two-seat reconnaissance planes. The compact C.XIV was the result.

Powered by a 220 hp Benz Bz.IVa engine, the C.XIV had little of the elegance of the larger C.XII and was the first Albatros aircraft with staggered wings, a feature intended to improve the field of view for the crew.

The C.XIV remained a prototype but had sufficient promise that it was developed into the C.XV reconnaissance aircraft, the last Albatros wartime two-seater to reach production.

Above: The Albatros C.XIV prototype shows its compact lines.

Below: This front view of the Albatros C.XIV prototype shows great attention to minimizing frontal area.

Albatros C.XV

Above: Albatros C.XV 7801/18 at Adlershof (Johannisthal?) was developed from the C.XIV prototype as the last Albatros C-type placed in production. Aerodynamically balanced ailerons improved maneuverability.

A redesign of the compact C.XIV prototype led to the production C.XV. Like the C.XIV, the C.XV was powered by a 220 hp Benz Bz.IVa engine. The C.XV closely followed the overall design of the C.XIV; however, a number of improvements were made to the design to improve performance and handling.

The C.XV was slightly larger than the C.XIV, but despite that was lighter, upgrading performance and maneuverability. To further improve handling and maneuverability, the ailerons were aerodynamically balanced, which reduced the force the pilot need to exert on the ailerons.

The crew members were placed close together to enable them to better coordinate during flight. To improve the pilot's field of view the upper wing cutout for the C.XV was enlarged compared to that of the C.XIV. These refinements of the basic C.XIV design created a fine, general-purpose reconnaissance aircraft that became the last Albatros wartime two-seater to reach production.

Few C.XV aircraft had reached the front before the Armistice, but the type went on to a successful post-war career both in civil and military use. Poland received 18, mostly new, Albatros C.XV biplanes obtained by Capt. William's mission in Gdansk. The aircraft were collected at Poznan-Ławica airfield and delivered to CWL and CSL for assembly. The first (CWL refurbished 23.1) was delivered to the 16th EW but crashed on 12 August 1920, soon after leaving the Warszawa-Mokotów airfield. It went into a spin from about 100 m and did not recover, the crew being killed in the crash. Two were reported to have served with the 12 EW against actions against Budyonny's cavalry. The type also served with the 8, 14, and 16 EW and the 21 EN, participating in the Battle of Niemen, actions in Groodno and Lida. After the war they were quickly dropped from frontline service and used mainly for training. The last C.XV was in service at Grudziadz until November 1921. One, marked D.414, served with the 21st Assault Squadron. The C.XV remained in service till at least 1926.

Lithuania operated one, serial N8, until at least 1927, when it appeared in a fish scale finish.

The German forces retreating from the Baltic in 1918 left behind a number of aircraft. From these Latvia acquired C.XV 7818/18, however it was considered too badly damaged to repair.

In 1922, the Nationalist Forces in Turkey sent Saffet Arikan and Nuri Conker to Germany to purchase military aircraft. 21 Albatros C.XV biplanes were obtained; however, it was difficult to ship them to Turkey due to the opposition and the Allied occupying forces. The aircraft were shipped via train to Russia and from thereby ship to Samsun, a Nationalist Turkish harbour on the Black Sea. On 29 July1922, Capt. Sadettin was sent with a pilot, technician, and a mechanic to prepare as many aircraft as possible and get them to Ankara. Only two Albatros C.XV machines were able to be erected from the remains of those shipped.

These two Albatros reconnaissance machines took part in the Turkish Great Offensive that commenced on 26 August

Above & Below: C.XV 7801/18 at the factory. All controls were aerodynamically balanced for improved maneuverability.

Above: C.XV 7801/18 front view illustrates the clean lines for a biplane.

1922. Their service life was brief, being struck off in 1923.

A photograph shows the wreck of a crashed C.XV in the markings of the Count Keller Air Detachment of the Western White Russian Army. The aircraft is thought to have served in the Riga, Latvia, theatre circa 1919.

The Kingdom of Albania apparently received five Albatros L.45, the C.XV civilian conversion. A number of C.XV biplanes entered the German civil register as Albatros L.47 civil conversions. There was also an L.47b, but the difference in this sub-type is unknown. D-109, D-140, D-179 , D-185, D-186, D-300, D-407 and D-586 are recorded as C.XV biplanes in different references. (Ries gives D-109 as a Rumpler D.I and D-140 as an Alb. C.III. D-179 is also recorded as Fokker D.VII 7609/18 used by the DLR company. D-185 is also recorded as a Sablatnig type and D-407 as a DFLW D VIIIa.)

Lt. G. Koppen of the Netherlands obtained Albatros C.XV 7838/18 as a civil aircraft with German registration D-407. He intended to fly this machine to the Netherlands East Indies in February 1920, but bureaucratic hurdles could not be surmounted and the flight was not attempted. In 1927 Koppen finally made a flight to Batavia as co-pilot to KLM pilot G. Frijns in a Fokker VIIa.3m.

Note: Postwar usage information courtesy of Colin Owers.

Above: Colorfully-marked Albatros C.XV number 8 in Lithuanian service gets inspected by the brass. (Peter M. Grosz Collection/SDTB)

Above & Below: C.XV 7801/18 at the factory. The front view shows the radiator that looks like the type applied to the Albatros J.II

Above & Below: C.XV with experimental dual radiators.

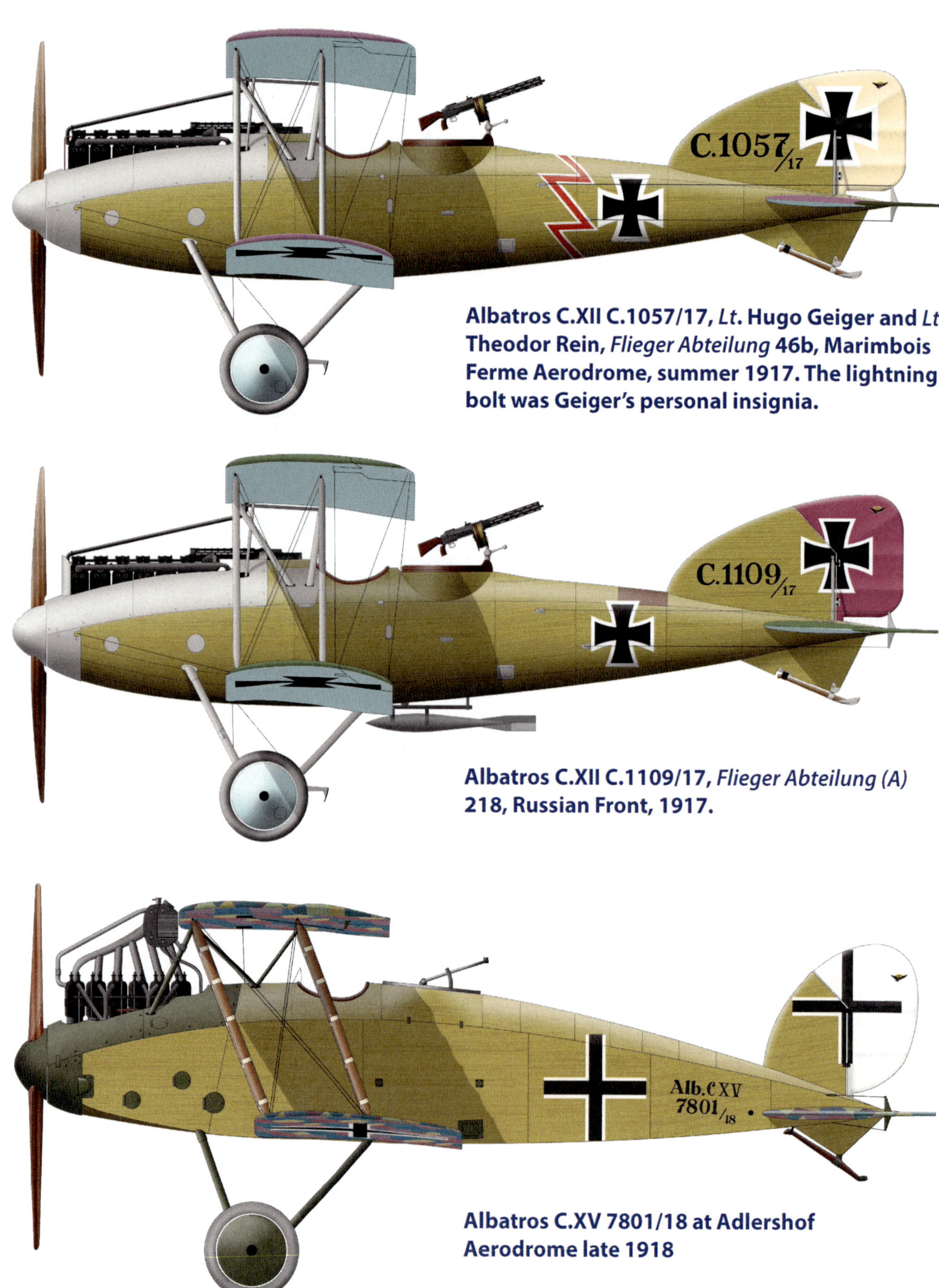

Albatros C.XII C.1057/17, *Lt.* Hugo Geiger and *Lt.* Theodor Rein, *Flieger Abteilung* 46b, Marimbois Ferme Aerodrome, summer 1917. The lightning bolt was Geiger's personal insignia.

Albatros C.XII C.1109/17, *Flieger Abteilung (A)* 218, Russian Front, 1917.

Albatros C.XV 7801/18 at Adlershof Aerodrome late 1918

Aviatik Introduction

Right: An Aviatik B-type, company designation P13, is in the foreground of this pre-war image of a flying competition. Aviatik soon earned a reputation for robust, reliable biplanes and 101 Aviatiks were ordered in 1913, a substantial quantity for this early period before the war. Italy even produced the Aviatik under license; the resulting SAML S.2, armed and using Italian engines, served until the Armistice.

The Automobil und Aviatik AG located in Mühlhausen-Burzweiler (now Mulhouse-Burzweiler, Alsace, France) was established on 10 December 1909 by combining the Fahrrad und Automobilfabrik with Aviatik GmbH. The new company began production by building Farman pusher biplanes and Hanriot monoplanes under license. In 1910 Aviatik delivered a biplane to the German *Fliegertruppe*, and in 1911 the *Fliegertruppe* ordered three modified Farman biplanes from Aviatik, followed by two monoplanes and nine biplanes ordered in 1912.

To achieve more orders and eliminate license production fees, Aviatik determined to design and build its own designs. *Diplom-Ingenieur* Robert Wild, Aviatik chief engineer, responded by designing a large, robust biplane that became the prototype for Aviatik's B-type aircraft. With this new design to offer, which had a good payload, a total of 101 aircraft were ordered from Aviatik in 1913.

Aviatik pilots Viktor Stoeffler and Carl Ingold established several pre-war distance and duration records that gave the Aviatik biplanes a solid reputation for reliability and performance. As a result, the Wild-designed Aviatik B and C-types were built in many varieties and were purchased in substantial numbers from 1913 through 1917. Aviatik biplanes were also manufactured under license by Weiser & Sohn in Vienna. Moreover, the Officine Moncenisio and Stablimenti Farina (SAML) in Italy and the Anatra company in Odessa built unauthorized variations of Aviatik B-type aircraft.

The Aviatik factory was close to the French border, making it vulnerable in case of war. However, Aviatik was prepared, and the factory was quickly relocated on mobilization day. With the French army dangerously close, the machine-tool park and production inventory was evacuated to Freiburg in an overnight transport comprising 50 freight wagons. According to Siegert's diaries, the evacuation took place on the first mobilization day, August 1 1914. Unfortunately, during the First Battle of Mühlhausen on August 12, Aviatik director Georg Chatel and his wife were killed by a stray artillery shell, proving the relocation necessary. Apparently production was hardly affected by the move. However, because Freiburg was within the range of Allied bombers, Aviatik was again forced to move, this time to Leipzig-Heiterblick, where production began in June 1916 in buildings designed and erected specifically for the Aviatik aircraft factory.

In July 1914, Aviatik formed a joint venture with A. Weiser und Sohn AG of Vienna. The new company was known as the Oesterreichisch-Ungarische Flugzeugfabrik "Aviatik" GmbH and was intended to supply aircraft to the Austro-Hungarian *Luftfahrtruppe*. The story of that venture and its aircraft are related in detail in *Austro-Hungarian Army Aircraft of World War One* by Peter M. Grosz, George Haddow, and Peter Schiemer and will not be repeated here.

Aviatik was a major aircraft manufacturer during the war and in addition to its own designs built other types under license.

Frontbestand Inventory of Aviatik Aircraft at the Front

Aviatik Type	1914			1915						1916						1917					
	31 Aug	31 Oct	31 Dec	28 Feb	30 Apr	30 Jun	31 Aug	31 Oct	31 Dec	28 Feb	30 Apr	30 Jun	31 Aug	31 Oct	31 Dec	28 Feb	30 Apr	30 Jun	31 Aug	31 Oct	31 Dec
B/12	1																				
B/13	34	24	13	4	4	2	1	1			1	1									
B.				17	23																
B.I	12	43	51	28	53	48	35	20	11		5										
B.II						1	29	38	32	18	7	2			1						
B.III										2											
C.I					1	8	9	35	81	161	150	219	183	177	110	63	27	12	1	1	
C.II								3		7	6				22	62	50	34	4	1	
C.III													26	47	22	20	6	2			
C.V																		1			
Total	37	67	64	49	81	59	74	97	124	188	169	222	209	224	155	145	83	49	5	2	—

Above: Aviatik B- and C-types were prominent at the front until early 1917, and as a result Aviatik had a strong company reputation. After newer Aviatik designs failed to achieve production, Aviatik, with its extensive production facilities, produced other company's designs under license. Some Aviatik aircraft were used for training in 1918 but none were at the front.

Aviatik C-Type Specifications

	Aviatik C.I	Aviatik C.II	Aviatik C.III
Engine:	150 hp Benz Bz.III 160 hp Mercedes D.III	200 hp Benz IV	160 hp Mercedes D.III
Span, Upper:	12.5 m (41.0 ft.)	11.71 m (38.4 ft.)	11.8 m (38.7 ft.)
Span, Lower:	10.75 m (35.3 ft.)	11.71 m (38.4 ft.)	10.2 m (33.5 ft.)
Chord, Upper:	1.88 m (6.17 ft.)	1.70 m (5.58 ft.)	1.70 m (5.58 ft.)
Chord, Lower:	1.88 m (6.17 ft.)	1.70 m (5.58 ft.)	1.70 m (5.58 ft.)
Length:	7.91 m (25.95 ft.)	6.95 m (22.8 ft.)	8.08 m (26.5 ft.)
Track:	1.94 m (6.36 ft.)	—	1.95 m (6.40 ft.)
Gap:	2.0 m (6.56 ft.)	1.80 m (5.91 ft.)	1.80 m (5.91 ft.)
Wing Area:	43 sq. m. (463 sq. ft.)	37.9 sq. m. (408 sq. ft.)	35.0 sq. m. (377 sq. ft.)
Empty Weight:	745 kg. (1,642 lb.)	916 kg. (2,019 lb.)	895 kg. (1,973 lb.)
Loaded Weight:	1,245 kg. (2,745 lb.)	1,509 kg. (3,327 lb.)	1,220 kg. (2,690 lb.)
Maximum Speed:	120 km/h (74.6 mph)	155 km/h (96.3 mph)	155 km/h (96.3 mph)
Cruise Speed:	—	—	140 km/h (87 mph)
Climb to 1,000 m:	5 minutes	—	6.5 minutes
Climb to 2,000 m:	16 minutes	—	16 minutes
Climb to 3,000 m:	33 minutes	—	31 minutes
Climb to 4,500 m:	—	—	55 minutes
Service Ceiling:	—	—	4,500 m
Duration:	3 hours	—	Range: 480 km
Armament:	1–2 flexible machine guns	1–2 flexible machine guns	1–2 flexible machine guns
Note: The C.III had 1 degree of dihedral on both upper and lower wings.			

Above: The specifications of the production Aviatik C-types are given in one table for ease of comparison.

Aviatik C.I

Above: The Aviatik C.I, developed from the earlier Aviatik B.II, and powered by engines of 150–160 hp, was one of the most numerous early C-types. At the dawn of air combat it was not yet certain that the gunner should be in the rear cockpit for a better field of fire to defend against attacking fighters, and *Idflieg* directed Aviatik to retain the observer in the front cockpit for better visibility forward and downward. A gun was mounted on each of the two rails alongside the front cockpit for the observer's use. This portrait is of Aviatik C.I 3522/15 and its ground crew.

The Aviatik C.I (company designation P25) was an armed development of the earlier Aviatik Type P15. It looked very similar to its predecessor and the airframe remained conventional wood, wire, and fabric. To improve performance while carrying the additional weight of machine guns, Aviatik engineers slightly enlarged the airframe and installed a more powerful 160 hp Mercedes D.III engine, then just entering quantity production. Machine-gun rails were fitted to both sides of the front cockpit and the wing bracing wires were moved to provide a greater forward field of fire for the guns. In mid-1915 *Idflieg* had not yet determined which layout – the machine-gun at the front or the rear – would provide the best all-round defense. To determine the superior setup, *Idflieg* ordered Aviatik C.I machines with the machine-gun armament placed in the front location, whereas Albatros C.I and LVG C.I aircraft were armed with a rear turret. No synchronized gun for the pilot was installed.

Airmen accustomed to the stable, sluggish B-types found that greater flying skill was required to handle the increased maneuverability and livelier control response of the new, more powerful C-types. However, improved performance and maneuverability were required by the increasing level of air combat and aircrews had to adjust. In mid-1915 the intensity and frequency of air combat were still relatively low, so the forward machine-gun location was not viewed as a particular disadvantage. *Feld-Flieger Abteilung* 6b reported that the field of fire from the front cockpit was 'medium good' and that successful attack or defense was primarily a function of the pilot's skill in air combat maneuvers. The Aviatik C.I arrived at the front in mid-1915 and served in quantity throughout 1916 before being replaced at the front in the first half of 1917.

Ironically, in April 1917, *Idflieg* ordered 50 Aviatik C.I biplanes fitted with a rotating turret behind the pilot for gunnery training. This is difficult to understand because C.I production had ceased in 1916 and Aviatik was fully engaged in production of DFW C.V(Av) biplanes at the time. Under the circumstances, this production program was accorded low priority; the first aircraft was not delivered until October 1917 and the type-test was not completed until January 1918. Aviatik records show that Leipzig-Heiterblick delivered one prototype and 40 production trainers powered by the 150-hp Benz Bz.III engine and Bork delivered ten C.I trainers powered by the 160-hp Mercedes D.III engine.

In-flight view of a well-worn Aviatik C.I; as shown by its boxy lines the C.I was a workhorse, not a thoroughbred.

Aviatik C.I(Han)

When the Hannoversche Waggonfabrik began aircraft production in 1915, the designation of the license-built Aviatik C.I originally assigned by *Idflieg* was Hannover C.I. However, this confusing designation was soon changed to Aviatik C.I(Han). The first example, C.1951/15 (w/n 1), was type-tested in February 1916. Hannover delivered a total of 146 Aviatik C.I(Han) aircraft to the *Fliegertruppe*.

Aviatik C.Ia

According to Aviatik, the Aviatik C.Ia biplane 'brought about several experimental improvements that would come into their full importance with the Aviatik C.III.' Aviatik engineers designed the C.Ia biplane to 'achieve greater airspeed at the expense of the useful load capacity' (the C.III had a slightly smaller wing area than the C.I). As such, the C.Ia, powered by a 160 hp Mercedes D.III engine, represented the first prototype for the Aviatik C.III, although the production C.III airframe would be further refined. The C.Ia prototype, ordered in April 1916 – the first in an order of 25 Aviatik C.III reconnaissance biplanes – probably made its appearance in May–June 1916. An undated crash photograph taken at *FEA* 3 shows that the aircraft was designated C.Ia 1750/16.

Left: Aviatik C.Ia 1750/16 was the aerodynamic test-bed and prototype for the Aviatik C.III powered by the same 160 hp Mercedes D.III engine as the C.I. The C.Ia gained a useful speed increase from its additional streamlining. The later Aviatik C.II also benefitted from these aerodynamic improvements although it used a different engine and airframe.

Above & Below: Early-production Aviatik C.I (or armed B) with side radiators; the observer is demonstrating his flexible machine gun (a captured Madsen?) while the excited pilot looks at the photographer. (Peter M. Grosz collection, STDB)

Above: Unit lineup of Aviatik C.I aircraft on a snowy field. (Peter M. Grosz collection, STDB)

Aviatik C.I 381/15 of a front-line unit with unit markings. A Fokker *Eindecker* escort is in the right background. (Peter M. Grosz collection, STDB)

Aviatik C.II

Above: The Aviatik C.II, developed from the C.I, used the more powerful 200 hp Benz Bz.IV engine, making it faster than the C.I. Unfortunately, at *Idflieg's* insistence it retained the observer in the front cockpit and the side-mounted guns, making for a cramped working environment and restricted field of fire. Unusual for a two-seater, the pilot had a headrest! Only a single production batch of 75 Aviatik C.IIs was constructed, after which Aviatik was directed to produce the DFW C.V under license. The Aviatik C.II and DFW C.V used the same engine and construction techniques, so had similar speed. However, the DFW had excellent maneuverability and handling qualities, critical attributes the Aviatik C.II lacked.

As the air war intensified and better opposing fighters appeared, faster C-types were needed to keep pace. The Aviatik C.II was developed from the earlier C.I in an attempt to provide the improved performance required. Like the C.I, it retained the awkward, observer-in-front seating arrangement and similar configuration and conventional construction. It was one of the first C-types to be powered by the 200 hp Benz Bz.IV, giving it significantly more power than the C.I. In addition to greater power, the C.II had a more streamlined nose than the C.I that included a propeller spinner, and the boxy, under-wing radiator of the C.I was replaced with a radiator of airfoil shape mounted in the upper wing. These aerodynamic improvements in the C.II were all based on the C.Ia prototype.

To further reduce drag, the wing span and area were reduced. A distinctive C.II feature was its vertical tail, which eliminated the fixed fin of the C.I in favor of a larger, strut-braced, horn-balanced rudder. Another distinctive C.II spotting feature was the pilot's headrest, which was unusual for a two-seater.

C.II Production and Service

In July 1916 an order was placed for 75 Aviatik C.II aircraft with 200 hp Benz Bz.IV engines, serials C.3075/16 to C.3149/16. An Aviatik C.II prototype (C.3685/16) was load tested between 23 September and 2 October 1916. The higher serial number (*Militärnummer*) for the prototype is explained by the fact, that prototypes – at least in 1916 – were bought or accepted by the authorities later than the series planes. Although later general practice was to order three prototypes for aircraft considered for production – a flying airframe, a static airframe for structural testing, and a spare – as far as is known only one C.II prototype was built. The date of the initial flight is unknown, so it is not known if the production order was placed before or after flight tests.

The Aviatik C.II appeared at the front in small numbers from late 1916 through mid 1917, a maximum of 62 being at the front at the end of February, 1917. The entries in the *Frontbeststand* table showing the C.II at the front before late 1916 are either transcription errors or a mix-up with license-builder designations, because the C.II production series was

Above: This side view of the Aviatik C.II shows its unique profile, emphasizing the small rudder without fixed vertical fin and the headrest for the pilot, who occupied the back seat. The type's handling limitations were designed in.

Below: This front quarter view of the Aviatik C.II shows the small rudder without fixed vertical fin and one of the gun mounting rails along the side of the observer's cockpit in front of the pilot. The radiator was now an airfoil type in the upper wing center section. In typical Aviatik fashion the wheel covers are marked with crosses.

not ordered until July 1916.

The Aviatik C.II was used on operations primarily, and perhaps solely, on the less demanding Eastern Front. Units known to have flown the C.II include *Flieger-Abteilung* 47b, *Flieger-Abteilung (A)* 244, and *Flieger-Abteilung (A)* 283. In addition, Aviatik C.II aircraft were noted at *AFP* 10, 12, and 14, all in the East, and later at various school units.

Right: Aviatik C.II preparing for a mission. There is a row of flares along side the observer's cockpit. (Greg VanWyngarden)

Left, Below Left, & Below-: An Aviatik C.II serving with *Flieger-Abteilung* 219 and its crew are shown in these views. The aircraft wears its factory finish with no unit or personal markings. The two machine guns for the observer, one on a rail on each side of his cockpit, are clearly shown, as are the metal rods that prevent him from shooting into his own propeller.

Right: Aviatik C.II has its portrait taken with one of its aircrew. The distinctive C.II identification features of side-mounted guns, metal rods to prevent the observer from shooting through his own propeller, and pilot's headrest are all clearly visible.

Left: Aviatik C.II on a typical snow-covered field on the Eastern Front. The rudder was painted in camouflage colors with a white-bordered iron cross insignia.

Below: An Aviatik C.II on a normal snow-covered field on the Eastern Front preparing for a mission. The under-size rudder with no fixed fin characteristic of the C.II shows clearly, as does the pilot's headrest

Aviatik C.III

Above: The Aviatik C.III, developed from the C.I, used the same 160 hp Mercedes D.III engine but was much more streamlined, making it faster than the C.I. Unfortunately, at *Idflieg's* insistence it retained the observer in the front cockpit and the side-mounted guns, making for a cramped working environment and restricted field of fire.

Like the Aviatik C.II, the Aviatik C.III was developed from the earlier C.I in an attempt to improve performance, especially speed. The C.III retained the same configuration, construction, and unhandy, observer-in-front seating arrangement as the C.I, and was powered by the same 160 hp Mercedes D.III engine. Like the C.Ia and C.II, the C.III had a more streamlined nose with propeller spinner, and the boxy under-wing radiator was replaced with one of airfoil shape mounted in the upper wing. Basically, the production C.III differed from the prototype C.Ia by having a smaller wing to reduce drag. Unfortunately, the smaller wing reduced payload and climb rate; giving the faster C.III basically the same climb rate as the C.I.

Unlike the C.II, the C.III retained the fixed fin and rudder of the earlier C.I and lacked the pilot's headrest that was unique to the C.II. To further increase speed, useful load was less than the C.I.

C.III Production and Service

The prototype for the C.III was Aviatik C.Ia C.1750/16 from the last C.I order of April 1916 for 75 Aviatik C.I (C.1700/16 to C.1774/16). The last 24 aircraft of that order (C.1751/16 to C.1774/16) were built as the C.III. A further 25 Aviatik C.III aircraft were ordered in July 1916 with serial numbers C.3150/16 to C.3174/16. In August 1917 there was a final Aviatik C.III order for 200 aircraft as trainers, serial numbers C.12201/17 to C.12399/17. These differed from the original C.III in having the pilot in the front and the observer, with his flexible machine gun, aft.

Interestingly, the Aviatik C.III preceded the C.II into service. Although the C.II and C.III shared many aerodynamic refinements from the C.Ia, the C.III was more closely related to the earlier C.I design and used the same 160-hp Mercedes D.III engine as the C.I. Furthermore, that engine was available earlier than the Benz Bz.IV used in the C.II.

Above: Based on its fuselage insgnia, Aviatik C.III 1753/16 apparently served with *Flieger-Abteilung (A)* 209 before being interned in the Netherlands, as shown by the over-painted national insignia. The pointed spinner differs from the more rounded spinner normally seen on Aviatik C.III aircraft.

Above: The Aviatik C.III had clean lines for a two-seater; this example has the rounded spinner normally seen on the C.III.

The C.III appeared at the front in small numbers from mid 1916 through early 1917, a maximum of 47 being at the front at the end of October, 1916. Operational units known to have used the Aviatik C.Ia/C.III include: *FA*25, *FFA*12, *FFA*42, *FFA*65, *FFA*67, *bFFA*4, *bFFA*6, *KS*25/*KG*5, and *FA*47b; school use was with *SchS*8.

Above: The Aviatik C.III 1753/16 after internment in the Netherlands. This aircraft was the only Aviatik to be interned by the Netherlands.(Colin Owers)

Right: An Aviatik C.III displays the clean lines that made it usefully faster than the C.I despite using the same engine, a 160 hp Mercedes D.III.

Aviatik C.I C.1428/15 assigned to *Flieger-Abteilung (A)* 206.

Aviatik C.I C.228/16 *LULU* in service with *Flieger Ersatz-Abteilung* 9.

Aviatik C.I 381/15
Possibly *Fl.Abt.(A)* 25

Aviatik C.II 3078/16, a typical production C.II.

Aviatik C.III
***Fl.Abt.(A)* 257**

Aviatik C.III serving with *Flieger-Abteilung* 25.

Aviatik C.IV

Above: The Aviatik C.IV two-seat reconnaissance plane prototype. The lineup of personalities above is an indication the aircraft may not be just an engine testbed but an attempt to develop a new production type. (Peter M. Grosz collection, STDB)

The Aviatik C.IV was the next Aviatik two-seater design. It was a two-seat reconnaissance plane prototype that appears to be basically a C.I fitted with a 200 hp Benz Bz.IV engine. The observer remained in the front cockpit and there was no fixed gun for the pilot. The aircraft remained a prototype.

Below: The Aviatik C.IV two-seat reconnaissance plane prototype. It appears to be basically a C.I fitted with a 200 hp Benz Bz.IV engine. As shown by the gun mounting rails, the observer was in the front cockpit. (Peter M. Grosz collection, STDB)

Above & Right: The Aviatik C.IV two-seat reconnaissance plane prototype with a 200 hp Benz Bz.IV engine. As shown by the gun mounting rails, the observer was still in the front cockpit. (Peter M. Grosz collection, STDB)

Aviatik C.V

Above & Below: The Aviatik C.V two-seat reconnaissance plane prototype. Powered by a 180 hp Argus As.III engine , the C.V was a complete break from the previous Aviatik types. It was a 'clean sheet' design whose configuration owed nothing to previous Aviaitik two-seaters. The wing cellule was completely new, featuring an interesting wing shape and strut design. The pilot was finally moved to the front cockpit and armed with a fixed, synchronized machine gun in the starboard side of the nose. A spinner was fitted and there was no fixed fin . (Peter M. Grosz collection, STDB)

The Aviatik C.V was the next true Aviatik two-seater design. Powered by a 180 hp Argus As.III, the C.V was of conventional construction but had an interesting gull-wing configuration to give maximum visibility and field of fire above the aircraft. The bracing struts were of significantly different design than earlier Aviatik C-types. The C.V differed from previous Aviatik C-types in having the pilot in the front cockpit with a fixed, synchronized Spandau and the observer in the rear cockpit with a flexible Parabellum. Some photos show a lip at the front of the aircraft where a propeller spinner was designed to be fitted; others show the aircraft with spinner. The C.V may have remained a single prototype, but one was shown at the front in June 1917 in the *Frontbestand* tables. This may have been an aircraft undergoing operational evaluation or could have been an error.

Above & Below: The Aviatik C.V two-seat reconnaissance plane prototype with spinner fitted for flight evaluation. There was no fixed fin. (Peter M. Grosz collection, STDB)

Left: This view of the Aviatik C.V two-seat reconnaissance plane prototype is the one that is usually published. The spinner was not fitted for this photo but was fitted for flight evaluation. (Peter M. Grosz collection, STDB)

Aviatik C.VI

Above & Below: Two views of the Aviatik C.VI two-seat reconnaissance plane prototype. The aircraft appears to be a rationalization of the C.V design with more conventional 2-bay wing cellule and fixed fin. The engine cowling was similar to the C.V but no spinner was used. Power was a 200 hp Benz Bz.IV. The Aviatik C.VI was Aviatik's unsuccessful attempt to design a better aircraft than the DFW C.V that it was then building under license. (Peter M. Grosz collection, STDB)

The Aviatik C.VI was a rationalization of the earlier C.V design. A conventional two-bay wing cellule was fitted, a fixed fin was added, and the propeller spinner was eliminated.

These changes resulted in a more conventional appearance and undoubtedly made the aircraft easier to build and fly. However, the C.VI was not produced in quantity, so this improvement to the Aviatik C.V still did not match the DFW C.V that Aviatik was then building under license.

Aviatik C.VII & C.VIII

Above: The Aviatik C.VIII was a small, single-bay biplane apparently designed as a light C-type. The C.VIII remained a prototype; the Halberstadt and Hannover designs were produced in quantity instead and both were very successful.

The Aviatik C.VII was apparently an un-built project, making the 1917 C.VIII the next Aviatik two-seater design actually built. The C.VIII was a small, single-bay biplane apparently designed as a light C-type in competition with the Halberstadt CL.II and Hannover CL.II. Like these competitors the C.VIII was powered by a 160 hp Mercedes D.III. To give sufficient gap between the wings to avoid aerodynamic interference, the lower wing was attached to a keel below the fuselage and braced with additional struts. The C.VIII remained a prototype.

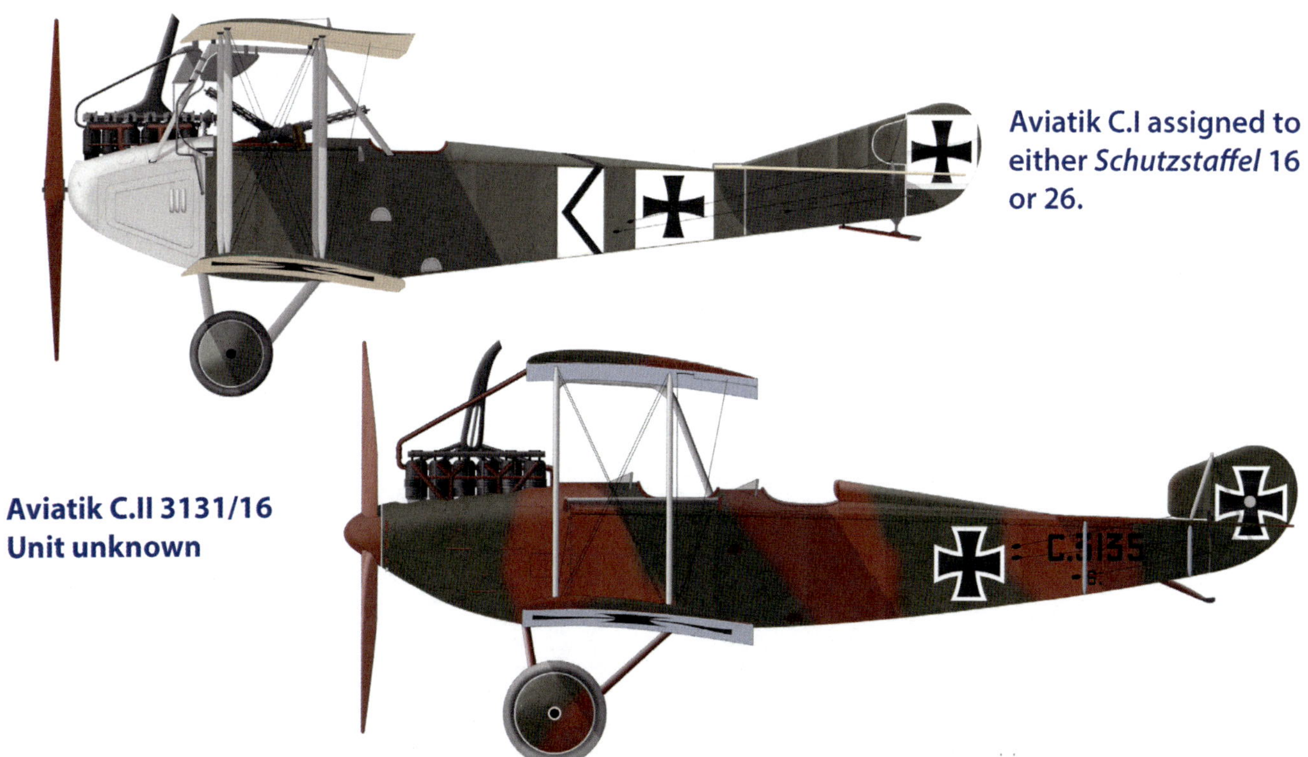

Aviatik C.I assigned to either *Schutzstaffel* 16 or 26.

Aviatik C.II 3131/16 Unit unknown

Aviatik C.IX

Above & Below: The Aviatik C.IX was the last Aviatik C-type built. The aircraft above is probably the first of three prototypes, while the aircraft below, which features an enlarged rudder and revised horn balances on the ailerons, is probably the second of three prototypes. Both aircraft have ailerons on the upper wing only.

Above: The Aviatik C.IX used the same 200 hp Benz Bz.IV engine used in the earlier C.II but was much more advanced aerodynamically. This aircraft, probably the third prototype of the three built, has ailerons on all four wings connected by actuating struts for higher roll rate and improved maneuverability. A nose radiator was used in the C.IX. The late-style crosses verify the date as sometime in 1918.

The C.IX was the final Aviatik two-seater design built. During 1918 three prototypes, 6306–6308/18, were constructed. Powered by the 200 hp Benz Bz.IV and mounting a nose radiator, the prototypes explored different tail surfaces and aileron configurations in an attempt to optimize maneuverability and flying qualities. The lower wing was suspended below the fuselage on struts for greater wing gap. The second airframe was tested to destruction at Adlershof in mid-June 1918. Armament was the then-standard fixed, synchronized Spandau for the pilot and flexible Parabellum for the observer. The Aviatik C.IX was not chosen for production.

Above: The Aviatik C.IX two-seat reconnaissance plane prototype. At least two different aileron designs are shown, one with horn balances. The aircraft was powered by 200 hp Benz Bz.IV and a nose radiator was fitted. The C.IX was the final Aviatik C-type design. The photo above is thought to be previously unpublished; poorer-quality versions of the other two photos were published in the book *Aviatik Aircraft of WWI* in this series. Three prototypes, serials 6306–6308/18, were built in 1918; 6308/18 is below. Although a handsome aircraft, the C.IX was not chosen for production. (Peter M. Grosz collection, STDB)

Aviatik F.I

Above: The Aviatik F, completed postwar as a transport, was apparently converted during production from the Aviatik F.I designed in response to *Idflieg's* new "F" class long-range reconnaissance aircraft. This is included as a reconnaissance plane that was formerly a C-type.

In September 1918 *Idflieg* defined a new aircraft class, "F", which stood for *Fernaufklärerflugzeuge* or long-range aerial reconnaissance aircraft. The classification was sent simultaneously to Rumpler, Halberstadt, L.V.G., Albatros, Aviatik, D.F.W., L.F.G. Roland, and Sablatnig, who were required to satisfy the following requirements:

- Maximum altitude (8,000 m).
- Speed at high altitude (160–170 km/h at 7,000 m, more than 140 km/h at 7,400 m).
- View for the pilot downward and to the sides, and the possibility to vertically see the ground.
- Observer space only for photographic equipment.
- Parachute.
- Fuel tanks for 325 gallons of gasoline, either a tank with two separate chambers or a double-walled, bullet-resistant tank.
- Wireless transmitter with heating.
- Liquid oxygen for breathing at altitude.
- A fixed forward-firing machine gun and a flexible machine gun in a rotating ring for the observer.

A number of engines were suggested, including the 350 hp Benz Bz.V V-12, the 350 hp Man V-10, the 260 hp Mercedes D.IVb, the 370 hp BMW V-12, and the 300 hp BuS.IVa six-cylinder. Examples of the Benz Bz.V V-12 were delivered in September. A Roland F.I to this requirement was ordered in September and delivered in December.

A competition between F-types similar to the fighter compeitions started in September but the Armistice brought the competition to a halt.

The aircraft that was completed postwar as the Aviatik F Limousine for two passengers and mail was almost certainly the Aviatik F.I long-range reconnaissance design converted to a transport. The Aviatik F was a large, three-bay biplane that bore a marked resemblence to the Aviatik D.VII fighter. The fact that a Benz Bz.V V-12 was installed, one of the engine types recommended for the F-class, and the fact that

Aviatik F.I Specifications

Engine:	350 hp Benz Bz.V (V-12)	
Wing:	Span	15.0 m
General:	Length	7.9 m
	Height	3.1 m
	Empty Weight	1,050 kg
	Loaded Weight	1,615 kg
Maximum Speed (at 2,000m):		175 kmh
Climb:	5000m	35 min
Ceiling (absolute):		6,500 m
Endurance:		3 hours
Range:		525 km

Above: An artist's rendering of the Aviatik F Limousine.

Above: Structure of the Aviatik F during completion as a Limousine carrying two passengers.

Top: The Aviatik F during flight evaluation.

Right: The Aviatik F during construction as an F.I.

the aircraft was completed in the first half of 1919 indicate that it was designed during 1918 for the F-class requirement. Furthermore, a photo of the aircraft under construction in the factory shows the rear cockpit ready for installation of a gun ring, while another photo of it outside the factory shows it now being built as a transport.

The Aviatik F was completed as a transport; whether it flew is not known for certain but is likely given that it was completed.

Note: Courtesy Reinhard Zankl.

Below: The Aviatik F during postwar flight evaluation. There are no records of it having flown, but that is highly likely considering that it was completed. Use of a two-bladed instead of a four-bladed propeller seems unusual considering the 350 hp delivered by the Benz Bz.V V-12.

Brandenburg Type DD & C.I

Above: The prototype Type DD (w/n DD94) was flown by Franz Reiterer to a new altitude record of 5,000 m in 58 minutes with three passengers on 22 September 1915. He broke two other records with the machine. *Flars* was so impressed they bought the machine. For military service Brandenburg installed a tubular scaffold that enabled a second gunner to aim his machine gun over the top wing while the rear gunner had his machine gun mounted on a spigot mount. This clumsy arrangement was not popular with crews and reduced performance. It was soon removed. *(AHT AL0579-050)*

The Brandenburg company had a complex, colorful history documented in *Hansa-Brandenburg Aircraft of WWI, Volume 1–Landplanes* by Colin Owers. In short, the German government required that the company, headquartered in Germany, was to supply Germany's needs first and only respond to Austro-Hungarian needs if it did not interfer with German requirements.

In early 1915 the German army grounded all Brandenburg airplanes because they were not structurally sound, and never ordered another airplane from Brandenburg. This was a major blow to the company; the German army ordered 95% of German warplanes and only

The prototype DD after purchase by Austria-Hungary.

5% were ordered for the German navy. And the modest industrial capacity of Austria limited their aircraft orders to a little more than 10% of German orders. This greatly limited the markets available to Brandenburg.

Regardless, Brandenburg achieved notable successes. The Brandenburg W.12 two-seat floatplane fighter was extremely successful. The W.12 made all previous German seaplane fighters obsolete and was the start of a long series of very successful types, such as the W.29 monoplane and the enlarged W.19 biplane and W.33 monoplane that offered greater range and endurance.

More significant to this volume, Brandenburg's Type DD was ordered by the Austro-Hungarian Army as the Brandenburg C.I. This sturdy airplane was very successful and with 1,258 C.I aircraft built of 4,768 accepted by the *Luftfahrtruppen* during the war; it comprised 26% of all army airplanes purchased by Austria. Its robust airframe accommodated increasingly powerful engines as they became available and it served successfully until the armistice. This is a remarkable record for a type designed in 1915.

Because Brandenburg was a German company, its aircraft are covered here despite its C.I serving exclusively with Austria.

Brandenburg C.I Production

Serials	Manufacturer	Notes
26.01 – 26.70	Phönix	160-hp AD
27.01 – 27.88	Phönix	160-hp AD
29.01 – 29.40	Phönix	200-hp AD
29.50 – 29.89	Phönix	200-hp Hiero
61.01 – 61.24	UFAG	160-hp AD
61.51 – 61.60	Brandenburg	160-hp AD
61.61 – 61.72	Brandenburg	160-hp AD
63.01 – 63.32	UFAG	160-hp AD
63.51 – 63.86	Brandenburg	160-hp Mercedes
64.01 – 64.48	UFAG	160-hp AD
64.51 – 64.72	Brandenburg	160-hp AD
67.01 – 67.56	UFAG	160-hp AD
68.01 – 68.64	UFAG	160-hp AD
69.01 – 69.48	UFAG	200-hp AD
69.50 – 69.99	UFAG	200-hp Hiero
129.01 – 129.72	Phönix	200-hp Hiero
169.01 – 169.22	UFAG	220-hp Benz
169.31 – 169.142	UFAG	220-hp Benz
169.151 – 169.180	UFAG	250-hp Benz
229.01 – 229.32	Phönix	200-hp Hiero
269.01 – 269.72	UFAG	200-hp AD
329.01 – 329.56	Phönix	200-hp Hiero
369.01 – 369.72	UFAG	230-hp Hiero
369.101 – 369.132	UFAG	230-hp Hiero
369.141 – 369.210	UFAG	230-hp Hiero
429.01 – 429.48	Phönix	230-Hiero

Note: AD = Austro-Daimler

Above & Right: *"Befehlsausgabe an die Flieger 28.6.18."* Pilots and observers of *Flik* 69S are briefed on their mission. The upper wing crosses have no white outline. Flik 69S specialized in ground attack missions.

Brandenburg-Built Brandenburg C.I Specifications

Source	Brandenburg 3-view Type DD	PMG Series 61.5	PMG Series 63.5	PMG Series 64.5
Span, m	12.300/11.700	12.30/11.70	12.30/11.70	12.30/11.70
Length, m	ca 8.150	8.15	8.40	8.30
Height, m	—	—	3.20	3.20
Sweepback	—	0°	0°	1°
Wing Area, m²	38.2	38.2	37.0	35.0
Wt. Empty, kg	741	741	770	745
Wt. Loaded, kg	1,215	1,215	1,219	1,219
Speed in km/hr	—	151	142	144
Time to 1000 m	6 min 40	5 min 20 sec	5 min 21 sec	6 min 35 sec
Time to 3000 m	—	25 min 10	—	—
Time to 5000 m		36 m 40	—	—
Motor	160-hp Austro	160-hp Daimler	160-hp Mercedes	160-hp Daimler

The source of the PMG specifications is *Austro-Hungarian Army Aircraft of World War One.*

Phönix-Built Brandenburg C.I Specifications

Source	PMG Series 26	PMG Series 27	PMG Series 29.5	PMG Series 129 & 229	PMG Series 29
Span, m	12.30/11.30	12.30/11.70	12.30/11.70	12.30/11.70	12.30/11.70
Length, m	—	8.40	8.37	—	8.37
Height, m	2.95	3.20	3.20	—	3.20
Sweepback	1°	1°	1°	1°	
Wing Area, m²	38.4	38.4	38.4	—	38.4
Wt. Empty, kg	797	778	805	794	811
Wt. Loaded, kg	1,238	1,219	1,238	1,230	1,260
Speed in km/hr	145	148	163	159	154
Time to 1000 m	6 min 20	5 min 30	4 min 12	3 min 58	4 min 50
Time to 3000 m	—	26 min 5	17 min 36	—	—
Time to 5000 m	—	—	54 min 40	—	—
Motor	160-hp Daimler	160-hp Daimler	200-hp Hiero	200-hp Hiero	200-hp Daimler

Left: Anton Riediger in Brandenburg C.I 169.119 of *Flik* 71D; the aircraft wears a colorful hexagonal camouflage scheme.

Above: This C.I has the Fischamend-designed MWF Type II upper wing gun installation with its disc-shaped ammunition canister under the wing that contains the ammunition belt for the over-wing gun.

Above: UFAG-built Brandenburg C.I 67.48 taking off from a typical front-line airfield showing the mountainous terrain.

UFAG-Built Brandenburg C.I Specifications

Source	PMG Series 61	PMG Series 64	PMG Series 68	PMG Series 63
Span, m	12.30/11.70	12.30/11.70	12.30/11.70	12.30/11.70
Length, m	8.23	8.23	8.23	8.23
Height, m	2.95	2.95	3.41	3.41
Sweepback	1°	1°	—	1°
Wing Area, m^2	38.5	38.5	38.5	38.5
Wt. Empty, kg	803	771	797	794
Wt. Loaded, kg	1,256	1,213	1,244	1,247
Speed in km/hr	138	140	146	—
Time to 1000 m	5 min 40	6 min 55	5 min 28	6 min 58
Motor	160-hp Daimler	160-hp Daimler	160-hp Daimler	160-hp Mercedes

UFAG-Built Brandenburg C.I Specifications

Source	PMG Series 67	PMG Series 69	PMG Series 169	PMG Series 263	PMG Series 369
Span, m	12.30/11.70	12.30/11.70	12.30/11.70	12.30/11.70	12.30/11.70
Length, m	8.23	8.35	8.35	8.35	8.30
Height, m	3.41	3.33	3.33	3.33	3.15
Sweepback	0.5°				
Wing Area, m^2	38.5	36		36	36
Wt. Empty, kg	789	799	930	827	880
Wt. Loaded, kg	1,236	1,251	1,381	1,278	1,331
Speed in km/hr	154	177	160	158	165
Time to 1000 m	5 min 05	5 min 24	4 min 15	4 min 30	4 min
Time to 3000 m			16-18 min	18-19 min	14-16 min
Time to 5000 m			50-55 min	50-58 min	48-55 min
Motor	160-hp Daimler	200-hp Hiero	250-hp Benz (Mar)	200-hp Daimler	230-hp Hiero

C.I 369.43 wears plain finish with a marking that appears to be the red/white/red marking of *Flik* 12.

Above: "Baby Coffin" with two machine guns installed.

Above: The observer is taking on carrier pigeons to a C.I identified as 169.112 on 2 July 1918. Pigeons were an essential method of sending messages before wireless telephony was truly developed. The hexagon camouflage is well displayed. Points to note are the slim machine gun barrel from the "Baby Coffin" (twin-barrel machine gun?); the forward gun sight on the exhaust pipe, and gun ring details. (For another view of this machine see *Cross & Cockade International* Vol.18, No.2, P.79).

Below: C.I 169.112 (?) setting off.

Type KDD

Above & Right: The KDD in its original form. The pilot was positioned beneath the upper wing in order to make the machine as streamlined as possible, but without the visibility required for a fighting machine.

The KDD was a two-seat reconnaissance plane derived from the KD and retaining the star-strut arrangement. The biplane was of typical Brandenburg mixed construction. Armament was a single machine gun. The pilot had limited forward view being enclosed in his cockpit by the upper wing and engine cowling in an attempt to streamline the aircraft. The first prototype, 60.56, was not successful although it underwent considerable modifications, including three changes of wings.

A second machine, 60.57, was built that incorporated all the lessons learned from the first. It too was unsuccessful but survived the war before being destroyed.

KDD text courtesy Colin Owers.

Type KDD Specifications

Source	*Typenschau**	PMG Data 60.56
Span, m	11.26	11.27/10.32
Chord, m		1.62/1.62
Length, m		7.75
Empty Wt., kg		820 kg
Loaded Wt., kg		1,184 kg
Motor	160-hp Austro-Daimler	160-hp Daimler

* The *Typenschau* also records the KDD as the UFAG KDD with the same specifications.

Above: This view shows the enclosed pilot's cockpit with Cellon windows. The second machine, 60.57 had a long fin installed.

Above: The KDD serial 60.56 on 5 July 1917. The original machine had been modified with the decking on the fuselage being raised to fill the wing gap, the struts have been faired and the stabilizer increased in area.

Brandenburg C.I 29.54 of *Flik* 23. The red fuselage band was the *Flik* marking.

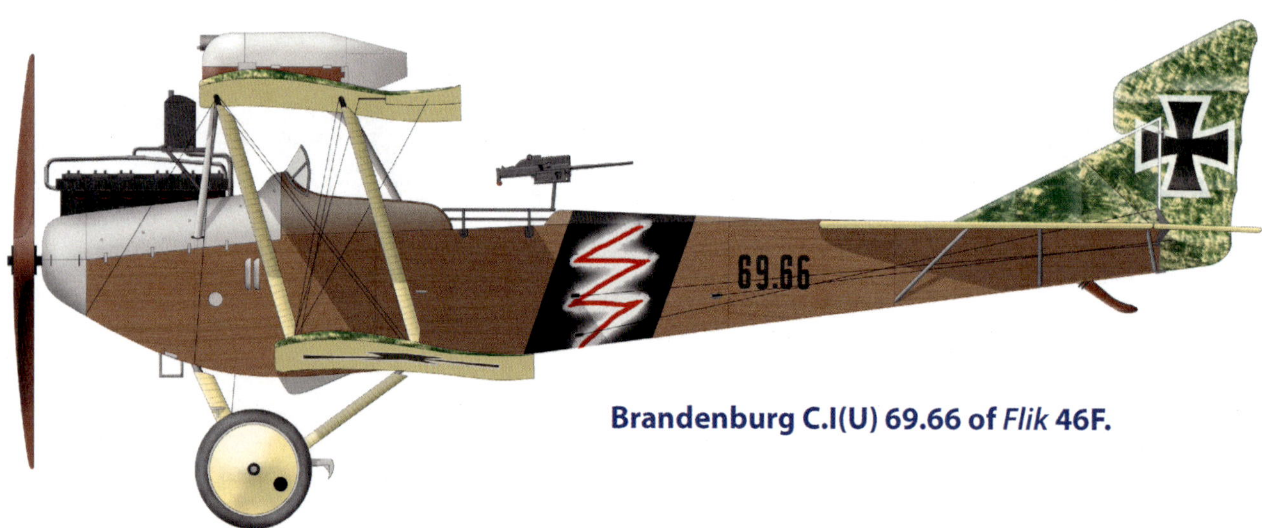

Brandenburg C.I(U) 69.66 of *Flik* 46F.

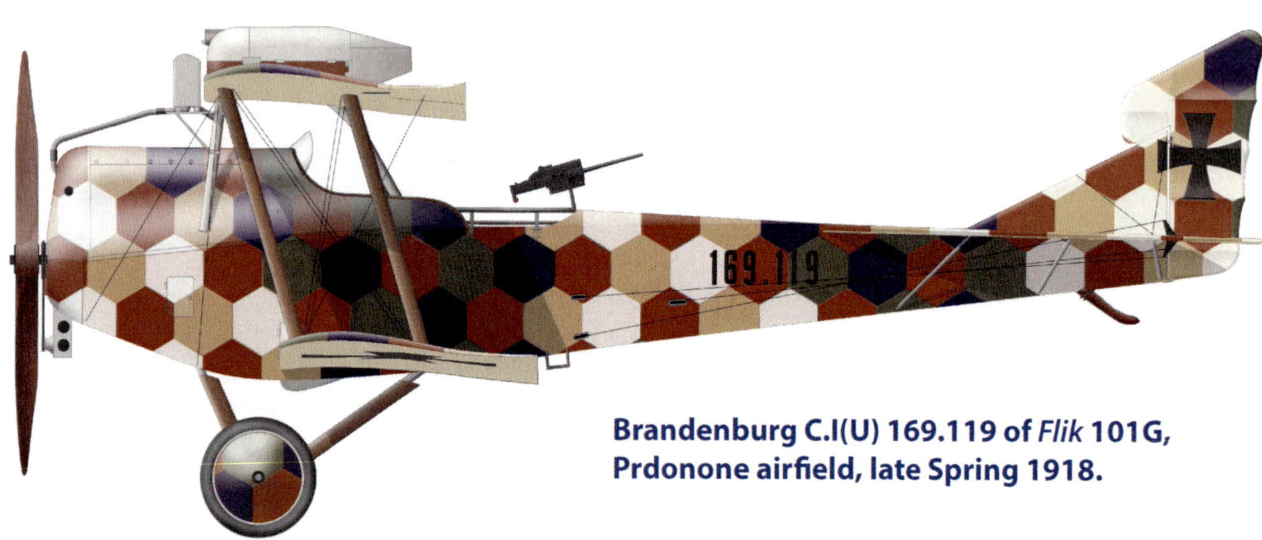

Brandenburg C.I(U) 169.119 of *Flik* 101G, Prdonone airfield, late Spring 1918.

Type K (C.II)

Above: The Type K/Brandenburg C.II 66.51 at Aspern during performance evaluation.

A two-seat observation biplane derived from the Type KDD the Type K became the C.II.[17] Two were constructed by Brandenburg. The machine had the star strutter form of interplane struts and as there were problems with the wing cellule the fuselage was constructed first while awaiting test results. The fuselage was deep and rectangular in section. The top wing was mounted close to the fuselage giving the observer a superb field of fire forwards and rearwards over the tail. A balanced rudder was fitted without fin. While the pilot was not enclosed as in the Type KDD, he was situated wholly under the upper wing with no easy escape in the event of a crash. A single machine gun was provided for the observer. On initial flight tests the performance was reported as good but Flars expressed concern over the pyramid strut arrangement. The prototype 66.51 was flown at Briest where it was inspected by Austro-Hungarian representatives. The pilot's view needed improvement and its performance was considered not satisfactory for the mountainous country that it would have to operate in.

The second machine, 66.81, carried out flight tests in June 1917 but had poor flight characteristics. Its poor longitudinal stability was blamed on the heavier 200-hp Hiero engine. Plans to lengthen the fuselage and provide an enlarged fin were halted while the L 15 was completed. By August 1917 the modified C.II was ready for flight test but the success of

Type K (C.II) Specifications

Source	*Typenschau* Type K	*Typenschau* UFAG KD*	Brandenburg 3-View	PMG 66.51
Span, m	9.40/9.40	9.40	9.400	9.40/9.40
Length, m	7.40	7.40	7.400	7.40
Wing Area, m²	28.83	28.83	28.83	28.03
Empty Wt., kg	736	736	728	728
Loaded Wt., kg	1,100	1,100	1,092	1,092
Time to 1000 m	3.5 min	3.5 min		3 min 30
Motor	180-hp Austro-Daimler	200-hp Austro-Daimler	180-hp Austro	185-hp Daimler

* The type is also identified in the *Typenschau* as the Phönix C.II with the same specifications.

the Phönix C.I and UFAG C.I reconnaissance two-seaters meant that the project was abandoned. Both these machines were developed with conventional wing bracing from their Brandenburg predecessors to become very successful machines.

KDD text courtesy Colin Owers.

Above & Below: The C.II was a typical Brandenburg structure with deep ply covered fuselage. The pilot's position under the upper wing gave him poor visibility and lack of egress in the event of a bad landing.

Type L 15

Above: The modified L 15 at Briest, 9 August 1917. The aircraft had to be extensively rebuilt in July 1917, the fuselage had to be strengthened and new two-bay wings installed. The machine was powered by a 345-hp V-12 cylinder Austro-Daimler engine.

The two-seat L 15 was Brandenburg's reply to a *Flars* specification for an aircraft powered by the new 345-hp Daimler engine. The L 15 was a large, good-looking two-seat machine with its propeller faired by a spinner into the neatly cowled V-12 engine. The pilot was equipped with a fixed, synchronized machine gun, and the observer had the usual flexible machine gun. Appearing in June 1917 the machine was nose heavy and broke its back in a hard landing. Rebuilt with two-bay wings, it survived the war.

L 15 text courtesy Colin Owers.

Type L 15 Specifications

Source	*Typen-schau*	Brandenburg 3-View	PMG data 2nd version
Span, m	13.80	13.800/ 13.800	13.80/13.80
Length, m	8.74	8.745	
Height, m			
Chord, m		2.250	2.25
Wing Area	52.66 m²		
Empty Wt.	1,015 kg	1,015 kg	1,015 kg
Loaded Wt.	1,500 kg	1,500 kg	1,500 kg
Motor	350-hp AD*	300-hp AD* V-12	345-hp AD*

* AD = Austro-Daimler

Below: Front view of the two-bay version of the L 15. The central airfoil-section radiator and the gravity tank may be clearly seen.

DFW C-Types

Above: Introduced in late 1916, the DFW C.V was still in front-line service in 1918. Although the most common aircraft in *Luftstreitkrafte* service, the example above, likely C.5967/17, is a rare night-bomber conversion, one of probably only four C.Vs thought to have been modified for this role. Six 50 kg P.u.W. bombs are carried, the number required by the *Iflieg* specification for single-engine night bombers, later to be designated N-types. The modifications included small navigation lights on both forward outer wing struts and likely the tail, the wing covering between two ribs on the lower starboard wing removed for the bomb sight on the fuselage, and the bomb sight. The black tube behind the bomb sight may be a night vision bomb sight (tested by German bomber units during WWI). The circular object mounted on the cabane is an altimeter. On the inner right front interplane strut under the Morell air speed indicator may be an additional *Fernfahrtmesser* (compass) for night flying. One landing light was mounted under the front fuselage. The white thunderbolt fuselage insignia indicates operational use; in this case, the aircraft was assigned to *Bogohl* 5.

The DFW C.V was not just the most important DFW C-type, it was the iconic DFW. Built in far larger numbers than all other DFW designs combined, the C.V appeared at the front in late 1916 and from its arrival quickly became the backbone of the general purpose German two-seater units. The DFW C.V was such an excellent all-around aircraft, combining reliability with excellent handling qualities and maneuverability, that it was still the most numerous two-seater late in 1918.

In contrast to the excellent C.V, the other DFW C-types made no significant impact on the air war. The earlier production types, the C.I, C.II, and C.IV, were built and operated in very small numbers and made only a minor contribution to German aviation, being inferior to their competitors from Albatros and Rumpler. Their production orders and operational service was little-noted at the time, leaving only a sketchy record.

The later C.VI and C.VII did not see service. The C.VI was not placed in production and the C.VII arrived too late. What follows is the best information we have been able to uncover on these rare aircraft; hopefully more will turn up in the future.

Known DFW C.I, C.II, & C.IV Production Orders			
Type	**Qty**	**Serials**	**Notes**
C.I	29	C.1500–1528/15	Probably the full serial number range
C.I	11	C.1975–1985/15	Probably the full serial number range
C.II	12	C.4200–4211/15	
C.IV	6	C.2374–2379/16	Probably not the full serial number range
These serial numbers are based on secondary data like photographs and strength reports, not primary sources like *Idflieg* orders. The highest serial numbers given for each type are the highest known.			

DFW B & C-Type Specifications				
	DFW B.I	**DFW C.I**	**DFW C.II**	**DFW C.III**
Engine	100 hp Mercedes D.I	150 hp Benz Bz.III	150 hp Benz Bz.III 160 hp Mercedes D.III	150 hp Benz Bz.III
Span, Upper	14.0 m	14.0 m	—	—
Span, Lower	—	—	—	—
Chord, Upper	—	—	—	—
Chord, Lower	—	—	—	—
Gap	—	—	—	—
Wing Area	—	40 m^2	—	—
Length	8.4 m	8.4 m	—	—
Height	3.18 m	3.18 m	—	—
Empty Weight	650 kg	800 kg	725 kg	—
Loaded Weight	1,015 kg	1,140 kg	1,235 kg	—
Maximum Speed	120 km/h	130 km/h	140 km/h	—
Climb to 1,000m	—	—	—	—
Climb to 2,000m	—	—	—	—
Climb to 3,000m	—	—	—	—
Climb to 4,000m	—	—	—	—

DFW C-Type Specifications				
	DFW C.IV	**DFW C.V**	**DFW C.VI**	**DFW C.VII**
Engine	150 hp Benz Bz.III	200 hp Benz Bz.IV	220 hp Benz Bz.IVa	220 hp Benz Bz.IVa
Span, Upper	13.30 m	13.27 m	13.60 m	13.60 m
Span, Lower	—	12.80 m	—	—
Chord, Upper	—	1.75 m	—	—
Chord, Lower	—	1.75 m	—	—
Gap	—	1.80 m	—	—
Wing Area	42.2 m^2	41.52 m^2	—	38.0 m^2
Length	7.90 m	7.88 m	7.50 m	7.0 m
Height	3.3 m	3.26 m	—	2.8 m
Empty Weight	720 kg	1,027 kg	—	800 kg
Loaded Weight	1,230 kg	1,477 kg	—	1,230 kg
Maximum Speed	135 km/h	155 km/h	—	160 km/h
Climb to 1,000m	—	4.0 minutes	—	3.0 minutes
Climb to 2,000m	—	7.0 minutes	—	—
Climb to 3,000m	—	15.0 minutes	—	—
Climb to 4,000m	—	25.0 minutes	—	—
Climb to 5,000m	—	40.0 minutes	—	—

DFW C.I

Above: The DFW C.I was developed from the earlier DFW B-types and retained the pilot in the aft cockpit. The observer occupied the front cockpit and the curved steel tubes from the forward cabane struts to the fuselage were fitted to prevent the observer from firing through the propeller arc. The C.I was powered by a 150 hp Benz Bz.III. *Uffz.* F. Decker was photographed in front of this DFW C.I used as a trainer at *Flieger Ersatz Abteilung* 6. (Peter M. Grosz Collection/SDTB)

DFW's first two-seater designed from the outset to be armed was the C.I. The C.I appears to be derived from the DFW B.II; the C.I had a more conventional, rectangular wing planform that abandoned the banana shape of the B.I. The superstructure for the observer's flexible gun used in the earlier armed B-types was eliminated to reduce drag; instead, the gunner, still in the front seat, had a gun ring with flexible machine gun. Curved steel tubes fitted on both sides of the aircraft between the forward cabane struts and the fuselage blocked the gunner from firing through the propeller arc.

The engine installed in the C.I was the reliable 150 hp Benz Bz.III that was more powerful than the 100 hp Mercedes used in most DFW B-types. Initially side radiators were used, but later C.I aircraft used a more modern leading edge radiator that had lower drag than the side radiators. The leading edge radiator was also less prone to leaking and, in event of damage, offered more time before the cylinder heads were drained of water, enabling the engine to run a little longer after the water started leaking.

Two small batches of the DFW C.I are known, C.1500/15 – C.1528/15) (29 aircraft) and C.1975/15 – C.1985/15 (11 aircraft). Apparently fitting of leading edge radiators started in the first batch.

The late Peter M. Grosz indicated the DFW C.I was removed from the front due to weak wing spars. It is noted that three C.I aircraft completed the type test January 15, 1916, well after the C.I reached the front, and at a time when no C.I aircraft are shown in the *Frontbeststand* inventory. This may be the second type test for the DFW C.I, confirming that the strengthened wing spars were satisfactory. The C.I retuned to the front in small numbers a couple of months later, likely after the aircraft removed from the front were rebuilt with stronger wings.

Above: The DFW C.I with the observer's flexible machine gun mounted on the gun ring in the front cockpit and the curved steel tubes from the forward cabane struts to the fuselage fitted to prevent the observer from firing through the propeller arc. Side radiators are fitted. (Peter M. Grosz Collection/SDTB)

Below: DFW C.I C.1512/15 or C.1513/15; the curved steel tube from the forward port cabane strut to the fuselage fitted to prevent the observer from firing through the propeller arc is clearly visible. This aircraft has a leading edge radiator instead of the side radiators of earlier C.I aircraft.

Above: Pristine DFW C.I C.1508/15 shows its overall light coloring. (Peter M. Grosz Collection/SDTB)

Above: This unidentified DFW C.I carries identification streamers and differently-shaped steel guardrails to prevent the observer from firing through the propeller arc. The presence of a pilot without observer may indicate it is in training service or that the pilot is preparing for a training flight. (Peter M. Grosz Collection/SDTB)

DFW C.II

Above: The DFW C.II used the engine and airframe of the C.I but the pilot was moved to the front cockpit. The observer occupied the rear cockpit and had a flexible machine gun. The curved steel tubes from the forward cabane strut to the fuselage fitted to the C.I to prevent the observer from firing through the propeller arc was no longer needed on the C.II. A leading edge radiator was used in place of the side radiators used on the C.I aircraft. (Peter M. Grosz Collection/SDTB)

The DFW C.II was a simple development of the C.I. The observer, together with his flexible gun, was moved to the rear cockpit and the curved steel tubes fitted on both sides of the aircraft between the forward cabane struts and the fuselage on the C.I were eliminated as no longer necessary to block the gunner from firing through the propeller arc.

The engine installed in the C.II was the same reliable 150 hp Benz Bz.III that was installed in the C.I. The C.II also used the leading edge radiator that had been installed in later production C.I aircraft.

Known production of the DFW C.II was very small, a total of 12 aircraft with serials C.4200/15 – C.4211/15.

Based on photographs, a small number of DFW C.II reconnaissance planes are thought to have been used operationally, but documentation is not available and they do not appear in the *Frontbeststand* inventory, so their service must have been brief.

Above: Portrait of a proud pilot in his DFW C.II. (Peter M. Grosz Collection/SDTB)

Above: The DFW C.II used the engine and airframe of the C.I but the pilot was moved to the front cockpit and the observer occupied the rear cockpit and had a flexible machine gun. (Peter M. Grosz Collection/SDTB)

Below: DFW C.II C.4207/15 in the factory. The observer in the rear cockpit is the key identifying feature of the DFW C.II compared to the C.I. Also, the C.II did not use the side radiators used on early C.I aircraft.

Above: DFW C.II C.2733/15 in flight; interestingly, this serial does not appear in the lists. (Peter M. Grosz Collection/SDTB)

Above: DFW C.II C.4203/15 in flight. (Bruno Schmäling)

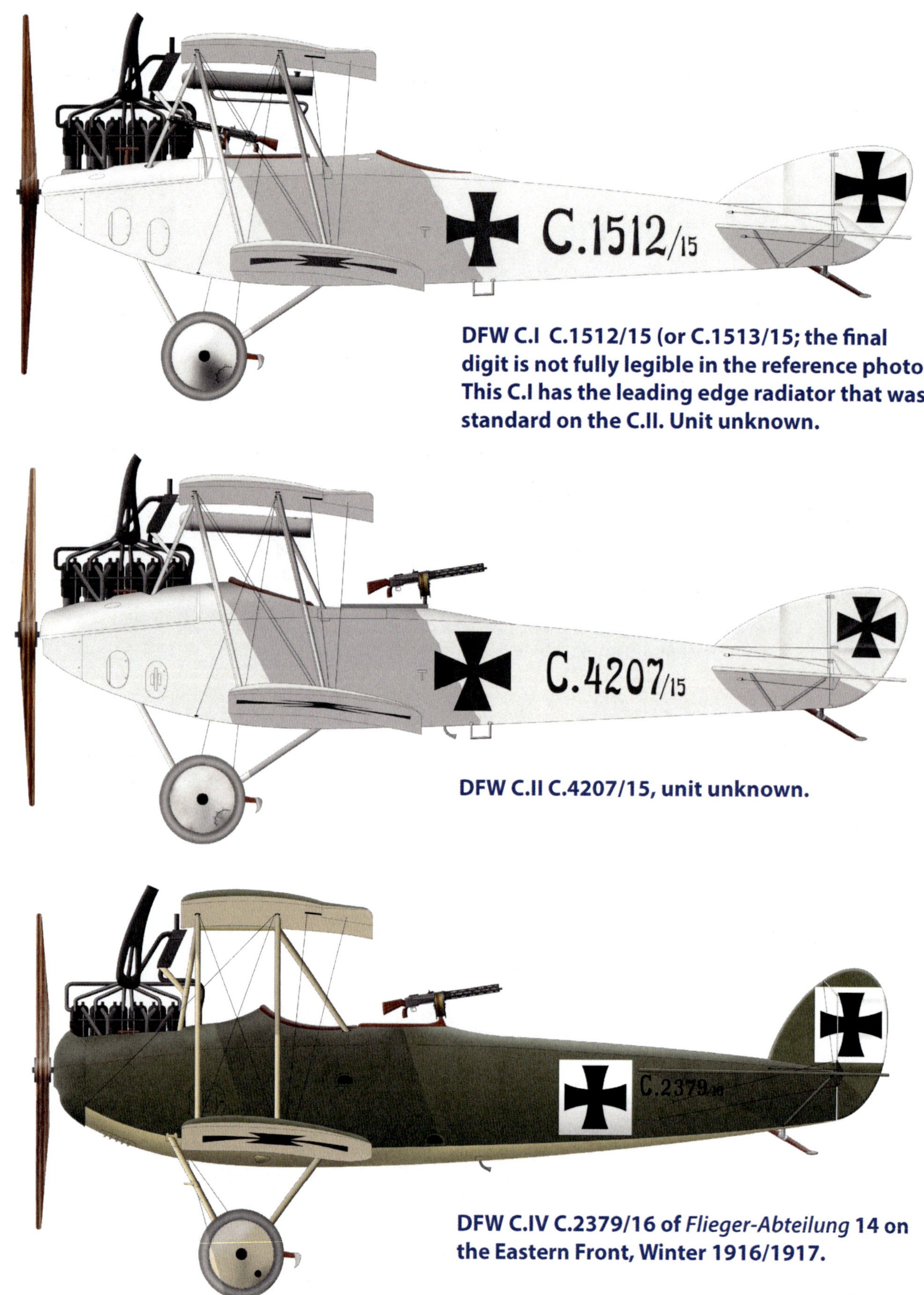

DFW C.I C.1512/15 (or C.1513/15; the final digit is not fully legible in the reference photo). This C.I has the leading edge radiator that was standard on the C.II. Unit unknown.

DFW C.II C.4207/15, unit unknown.

DFW C.IV C.2379/16 of *Flieger-Abteilung* 14 on the Eastern Front, Winter 1916/1917.

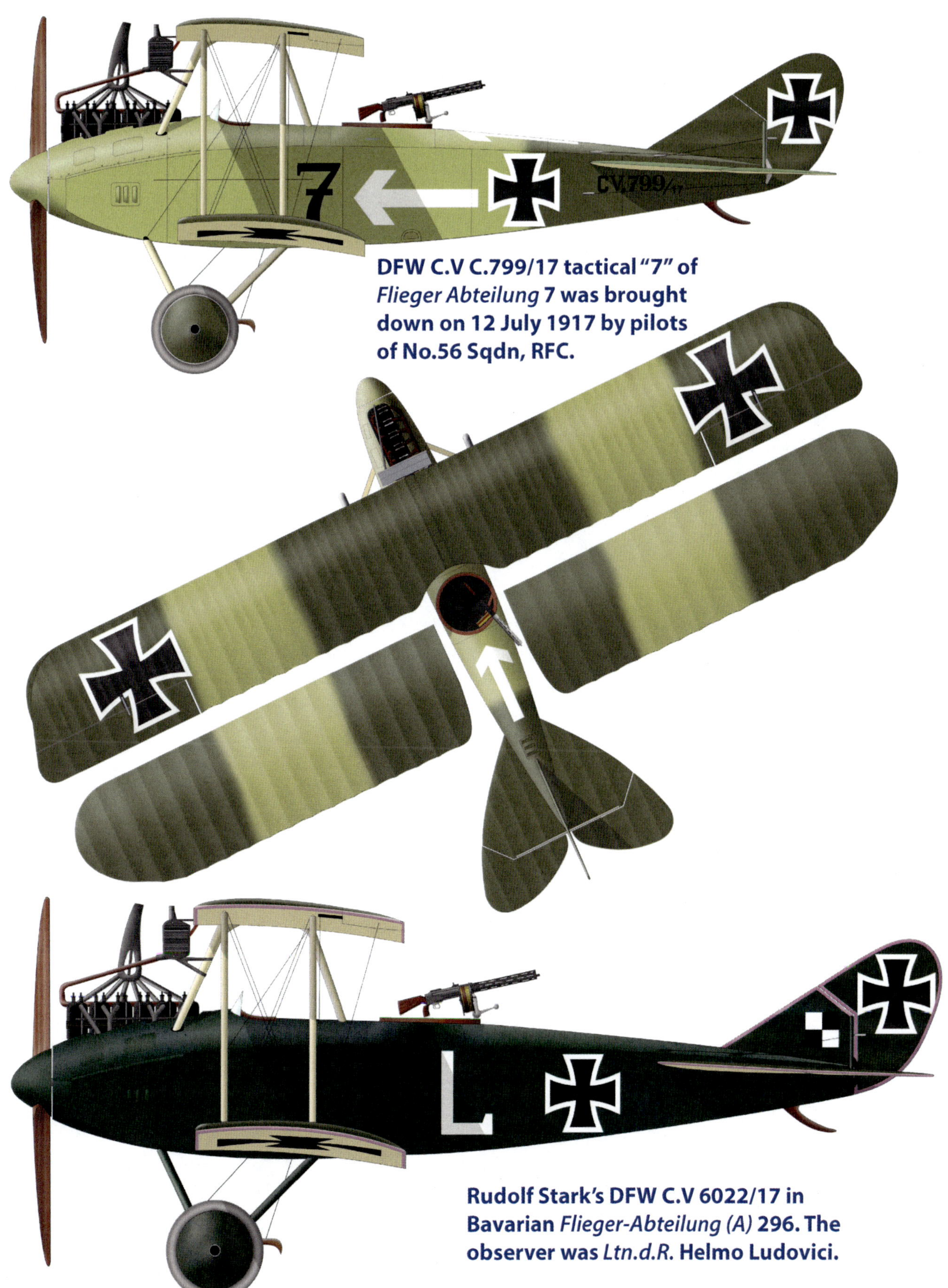

DFW C.V C.799/17 tactical "7" of *Flieger Abteilung* 7 was brought down on 12 July 1917 by pilots of No.56 Sqdn, RFC.

Rudolf Stark's DFW C.V 6022/17 in Bavarian *Flieger-Abteilung (A)* 296. The observer was *Ltn.d.R.* Helmo Ludovici.

DFW C.III

Above: This two-seat DFW pusher is thought to be the C.III but this is not confirmed. Like the C.I and C.II, the engine was a 150 hp Benz Bz.III. The pusher configuration was not popular in Germany due to its high drag, which limited performance and the observer's field of fire to the rear, increasing the difficulty in defending the aircraft from rear attack.

DFW built a prototype pusher two-seater shown in the above photograph. This aircraft, thought to be designated the DFW C.III, may have remained a single prototype but was photographed with two significantly different radiator arrangements so more than one prototype may have been built. The engine installed in the C.III was the same 150 hp Benz Bz.III that was used in the DFW C.I and C.II.

Pushers were not popular in Germany and few pusher designs were built. Although pushers had the advantage of unobstructed forward view and field of fire for the observer, the extensive structure supporting the tail created excessive drag, limiting speed and climb. Worse, a gunner in the front cockpit had a limited field of fire to the rear, and none to the rear and below, making pusher aircraft very vulnerable to fighter attack.

The C.III appears to have been built using the same engine as the tractor C.I and C.II to compare the two configurations. The limitations of the pusher arrangement clearly made the success of such an experiment unlikely, and no production of the pusher was undertaken.

Left: This view of the DFW C.III shows the 3-bay wing cellule to advantage. This aircraft has side radiators and was likely the first configuration of the C.III. (Peter M. Grosz Collection/ SDTB)

DFW C.IV

Above: With its rounded plywood fuselage and single-bay bracing the DFW C.IV was a complete departure from earlier DFW C-types. However, the engine remained a 150 hp Benz Bz.III. A leading edge radiator was fitted. An identification streamer is attached to the lower left wing tip and a barograph is suspended between the wings. (Peter M. Grosz Collection/SDTB)

In an attempt to produce a faster two-seater using the same 150 hp Benz Bz.III fitted in the preceding DFW C-types, DFW developed a streamlined, single-bay reconnaissance aircraft. The factory designation for this aircraft was T 25 and its military designation was C.IV. DFW technical director *Diplom-Ingenieur* Hermann Dorner was also responsible for designing the DFW C.IV, "built without conventional conceptions and utilizing structural materials to their limit."

The fuselage of the DFW C.IV was made of a plywood skin covering a wood structure. This streamlined form, together with the single bay wing design, reduced drag compared to the DFW C.I and C.II. With their C.IV, DFW took a similar approach to improved performance via reduced drag that LFG did with the Roland C.II. However, the Roland C.II was a more radical design than the DFW C.IV; the Roland used the fuselage to completely fill the gap between the wings, eliminating the cabane struts, and used streamlined, single I-shaped interplane struts.

The C.IV was fitted with a flexible gun for the observer but apparently there was no fixed gun for the pilot as none of the available photos show a synchronized gun.

Only a very small number of DFW C.IV aircraft were built. Known serial numbers are C.2374–2379/16, for a total of six aircraft. However, it is thought that more than six aircraft were built.

Photos in this section show the C.IV in service with *Flieger-Abteilung* 14 and 15 on the Russian Front, proving that at least a small number of DFW C.IV aircraft saw operational service despite not appearing in the *Frontbeststand* chart for C-types. These charts were compiled after the war from other reports and contain a number of errors, apparently including the omission of the small numbers of the DFW C.IV at the front. From other documents we learn that at *AFA* 220 / *FA(A)* 220 *Lt.* Hans Pipart flew DFW C.IV 2378/16 during the period November–December 1916. Several C.IV aircraft flew with training units later in 1917 after the C.IV had been withdrawn from combat.

Above: Men of *Flieger Abteilung* 14 on the Eastern Front pose for their portrait with DFW C.IV C.2379/16. The damaged rudder likely indicates the aircraft over-turned on landing. (Peter M. Grosz Collection/SDTB)

Above: The full photograph shows a DFW C.IV on its way to the front during the winter on a pair of railroad cars, the fuselage on one car and the wings on another. The circle on the wing root is likely the magnetic compass housing.

The crew of a DFW C.IV of *Flieger Abteilung* 14 (or 15?) poses for their portrait. (Peter M. Grosz Collection/ SDTB)

A ground crewman warms up a DFW C.IV. The rounded plywood fuselage was carefully streamlined and well-finished to minimize drag. (Peter M. Grosz Collection/SDTB)

DFW C.V

Above: DFW C.V with *Lt. Graf* Fenster waving to the camera aircraft. This C.V was assigned to 1 *Marine Feld Flieger-Abteilung* and marked with a snake and light spinner; the pilot was *Flugmeister* Bartschis. The DFW C.V was the most numerous and important German two-seat reconnaissance aircraft of the war.

Without the DFW C.V, the DFW company would be a footnote in history as the producer of a variety of 'also ran' designs that sometimes supplemented the more successful designs with a few aircraft at the front. Of all DFW types, only the C.V was produced and operated in sufficient numbers to make a significant impact on the air war.

DFW hired *Diplom-Ingenieur* Hermann Dorner in July 1915 as technical director. Dorner, who had been working at the Adlershof *Prüfanstalt und Werft* (*PuW* – test establishment and workshop), devoted much of his effort to the DFW R-plane projects. However, he also designed the DFW C.III and C.IV two-seat reconnaissance planes. According to an *Iflieg* letter of July 19, 1916, the C.III airframe was not strong enough due to the I-strutted wing cellule. This indicates that the C.III may be the single-bay tractor biplane with I-struts shown on page 99, although the DFW pusher C-type is generally considered to be the DFW C.III. Dorner was also part of the DFW C.V design team.

The development history of the DFW C.V is obscure. Most likely the C.V design was the product of a team of DFW engineers, including Heinrich Oelerich, Hermann Dorner, and *Diplom-Ingenieur* Willi Sabersky-Müssigbrodt, who had joined DFW January 1, 1916 and led the C.V

Above: A prototype DFW C.V in front of the DFW factory. The first three prototype C.V aircraft are thought to have serials C.3300–3302/16, the first three serials from the first production batch of 60 aircraft ordered from DFW. The C.V was a sturdy, two-bay biplane powered by a 200 hp Benz Bz.IV. The prototypes and early production aircraft featured ear radiators; later production aircraft had a leading edge radiator for improved reliability and reduced drag.

Above: DFW C.V C.3311/16 was a C.V prototype or early production aircraft with ear radiators. The propeller lacks a spinner but the other C.V identification features are present. This photo became Sanke card 1042.

DFW C.V Production Orders					
Year	**Mon.**	**Manufacturer**	**Qty**	**Serials**	**Notes**
1916	Aug.	DFW	60	C.3300–3359/16	Includes prototypes 3300–3302/16
	Sep.	DFW	40	C.3400–3439/16	
	Oct.	DFW	240	C.4800–5039/16	
	Oct.	LVG	250	C.5040–5289/16	Type-test February 1917
	Oct.	Halberstadt	75	C.5750–5824/16	Type-test February 1917
	Oct.	Aviatik	150	C.5825–5974/16	Type-test February 1917
	Oct.	LVG	50	C.7650–7699/16	
	Oct.	BFW	300	—	Cancelled
	Nov.	Aviatik	100	C.9005–9104/16	
1917	Jan.	Aviatik	100	C.100–199/17	
	Jan.	DFW	150	C.750–899/17	
	Feb.	Aviatik	100	C.2000–2099/17	
	Feb.	LVG	100	C.2100–2199/17	
	Feb.	Halberstadt	25	C.2500–2524/17	
	Apr.	Halberstadt	50	C.3378–3427/17	
	Apr.	Aviatik	200	C.3800–3999/17	
	Apr.	DFW	150	C.4578–4727/17	
	May	DFW	200	C.6000–6199/17	
	Jul.	Aviatik	150	C.6900–7049/17	
	Jul.	DFW	200	C.7750_7949/17	
	Oct.	DFW	15	—	Type C.Vc, serial numbers unknown
1918	Jan.	DFW	300	C.17000–17299/17	Type C.Vc
	Feb.	Aviatik	150	C.200–349/18	
	Feb.	DFW	150	C.350–499/18	Some fitted with NAG engines
	Mar.	Aviatik	150	C.2150–2299/18	
	Mar.	DFW	150	C.2300–2449/18	
	Jun.	Aviatik	100	C.4300–4399/18	
	Jun.	DFW	100	C.4400–4499/18	
	Jun.	Aviatik	100	C.4500–4599/18	
	Jun.	Aviatik	50	C.6750–6799/18	
	Jul.	DFW	50	C.8000–8049/18	
	Jul.	Aviatik	200	C.8200–8249/18	
	Nov.	DFW	200	—	Type C.Vc, serials unknown
Summary: **3955** total aircraft ordered – DFW: 2005; Aviatik: 1400; LVG: 400, Halberstadt: 150					

Above: The DFW C.V was ordered into production in August 1916 and arrived at the front in late September/early October, when the Allied fighter opposition was composed of DH.2s and Nieuport 11s, 16, and 17s. Remarkably, the C.V was still being ordered in large numbers throughout 1918 when the SE.5a and the latest Spads were the predominant Allied fighters. The excellent maneuverability and handling qualities of the DFW C.V made it a tough opponent for even the best fighters and it was well liked by its crews. Newer German two-seaters like the LVG C.V and C.VI and Halberstadt C.V offered somewhat better performance than the DFW but never completely eclipsed it. In fact, LVG hired the designer of the DFW to create their LVG C.V to avoid having to produce more DFWs under license.

design team. The C.V, internal factory designation T 29, had a light but strong airframe due to careful design, and incorporated the new 200 hp Benz Bz.IV engine that offered a useful power increase over the 150 hp Benz Bz.III used in earlier DFW C-types. The C.V mounted the standard C-type armament of a flexible gun for the observer and a fixed, synchronized gun for the pilot; in the C.V the pilot's gun was on the right.

The C.V's chief designer, Willi Sabersky-Müssigbrodt, recalled during an interview that the first flight of a C.V prototype was in May 1916. On July 11 the *PuW* reported that the new "DFW C-type with 200 hp Benz Bz.IV still has so many shortcomings ... the factory must demonstrate a new aircraft. The flight performance is up to expectations." However, C.V prototype C.3302/16 passed the load tests a few weeks later, on August 3rd and 4th, and on August 6th *Idflieg* reported "the performance figures achieved by the DFW 200 hp are very good... Sixty aircraft were ordered."

Production started immediately, and the first C.V aircraft reached the front in late September/early October 1916, by which time an additional 40 aircraft had been ordered. The DFW C.V must have made an excellent impression very quickly because in October the amazing total of 1,065 C.V aircraft were ordered! This number was so large that four different license manufacturers, Aviatik, BFW, Halberstadt, and LVG, had to be brought into the DFW C.V production program to deliver the aircraft ordered. The BFW contract was later cancelled, but production examples from the other three license manufacturers all passed their type tests in early February 1917. Soon after that the inventory of DFW C.V aircraft at the front jumped to become the most numerous C-type at the front by the end of April, and by the end of August reached 1,057 aircraft! And despite introduction of newer designs, the C.V remained in production until the Armistice and remained the most numerous C-type at the front.

Why did the DFW C.V quickly replace the Aviatik C.II and other competing C-types in production? In particular, the Aviatik C.II was already in production using the same engine and structural technology as the DFW C.V; why take the time to change production over to the DFW when Aviatik already had a similar internal design in production?

The key to the success of the DFW C.V was its excellent handling and maneuverability combined with a robust structure able to endure rough use in the field. Although its modest speed and good climb were surpassed by later German C-types, the C.V was effective, reliable, and able to defend itself throughout the war.

By late 1917 Allied fighters certainly had a signiticant speed advantage over the C.V and could often intercept it, but that did not mean they could shoot it down. An impressive example occurred on 12 December 1917, when James McCudden intercepted a DFW C.V. McCudden was a leading British ace, likely the highest scoring British ace of the war in that more of his claims can be definitely confirmed by wrecked aircraft than any other British fighter pilot. He was eventually credited with 57 victories, and by the date of this combat had 45 confirmed victories.

In any case, McCudden was flying his SE.5a, the best all-around British fighter to see widespread service, and on this day sighted a DFW C.V returning to German lines at 4,000 feet, just under an overcast. Leaving the rest of his flight above the clouds to intercept the C.V if it tried to escape into the clouds, McCudden dived to the attack. But the experienced C.V crew did not try to escape; instead, they stayed down to fight McCudden. Their confidence in

Above: DFW C.V tactical number '7' is thought to be C.V C.4686/17 that was captured by the British and given captured aircraft number X.G.13. The white circles may be patched bullet holes. A leading edge radiator is fitted. A sample of the fabric from this aircraft is kept in the IWM; the deep green applied to the canvas gives a strong dark greenish-blue appearance.

themselves and their DFW C.V was justified; they fought McCudden for fully five minutes, but he simply could not gain a favorable attacking position. According to McCudden, "EA put up a most determined and skillful fight and I was not able to use his blind spots for a single second. Moreover, the enemy gunner was shooting very accurately and making splendid deflection, so that when I got down to 500 feet I left the EA who went north over Bourlon Wood." McCudden realized that he and his SE.5a had met his match in this particular two-seater crew and aircraft. "The Hun was too good for me and shot me about a lot. Had I persisted he certainly would have got me for there was not a trick he did not know."

This is a remarkable tribute to the DFW C.V and its crews from a remarkable adversary, and demonstrated that the DFW had the maneuverability and handling qualities to defend itself against the best Allied fighters. With such an excellent aircraft, combat results came down to the crew.

Understandably in light of its prolonged production, the C.V underwent some evolution during is manufacturing life. Early aircraft featured ear radiators, spinners, and full engine cowlings. The major change in production was replacement of the ear radiators with more effective leading edge radiators with lower drag and improved reliability. Different manufacturers applied somewhat different camouflage and markings, and pre-printed camouflage fabric was used on later aircraft. In the field the upper engine cowling was often removed during warm weather, and at times the propeller spinner was also omitted in hot weather for better cooling.

Finally, a training version, the DFW C.Vc, was developed. This version used the 185 hp NAG C.III engine that was a little longer than the Benz Bz.IV, requiring a lengthened cowling. The NAG became available in mid-1917 and a total of 515 C.Vc trainers were ordered from October 1917. The 'c' suffix in the designation stood for 'Conrad', designer of the NAG engine.

Right: Aviatik-built DFW C.V(Av) being spray-painted at the Aviatik factory. Aviatik built one batch of 75 of their competing C.II, than at least 1400 DFW C.Vs under licence in at least 13 batches. The aircraft used the same engine and similar technology. This is a testament to the excellent sturdiness and handling qualities of the DFW C.V.

Above: LVG C.VI 5040/16 (later rationally re-designated DFW C.V(LVG)) at Adlershof for its type test in February 1917.

Right: DFW C.V(LVG) 5055/16 tactical '4' of *Flieger-Abteilung (A)* 211. The dark fuselage band appears to be the unit marking. (Bruno Schmäling)

Above: DFW C.V 4430/18 of Bavarian *Flieger Abteilung (A)* 286 in May/June 1918. The camouflage is standard for this production batch with printed camouflage fabric on the flying surfaces and grayish-magenta and olive paint on the fuselage. The interesting symbol is copied from the logo of a Bavarian brewery and was the unit marking.

Left: DFW-built C.V tactical '2' of *Flieger-Abteilung (A)* 208; it has leading-edge radiators.

Above: Aviatik-built DFW C.V apparently belonging to *Flieger-Abteilung (A)* 209.

Above: Well-known photo of DFW C.V(Av) C.5845//16; the Aviatik logo was applied to the struts of Aviatik-built C.V aircraft and center of the rudder insignia as shown. Signal flares are in a bracket just behind the observer's cockpit.

DFW C.VI

Above & Below: The DFW C.VI was powered by a 220 hp Benz Bz.IVa. The control surfaces were horn-balanced to reduce control forces and improve maneuverability. It is thought only the single prototype was built.

DFW clearly wanted to follow the great success of their C.V with an improved design, and the C.VI was their next attempt. The C.VI used the slightly more powerful 220 hp Benz Bz.IVa for enhanced performance. To improve maneuverability, horn-balanced control surfaces were fitted which reduced the pilot's control forces to improve responsiveness, and the rudder and fin were enlarged.

Apparently the C.VI offered insufficient improvement over the successful C.V that was already in large-scale production and only a single prototype is thought to have been built and evaluated.

DFW C.VII

Above: The final DFW two-seater design to be completed had the internal company designation F 37; it is thought to have been assigned the military designation DFW C.VII but this is not confirmed. It was powered by a 220 hp Benz Bz.IVa. When fitted with a 260 hp BMW.IV engine postwar it achieved an altitude record of 7,700 m. (Peter M. Bowers Collection/ Museum of Flight)

The next and last DFW two-seater design was the F 37, thought to have been designated the C.VII. The C.VII used the 220 hp Benz Bz.IVa installed in the C.VI to improve on the performance of the C.V. The C.VII appears to have used the wing cellule tested in the C.VI although with the lower wingtips less rounded than those of the C.VI. The fuselage of the C.VII was more compact than the C.VI and a much larger rudder and fin were installed with different horizontal tail design, these changes being made to further improve maneuverability and handling qualities. In any case, the C.VII came too late to reach production or service.

Postwar a C.VII prototype was fitted with a 320 hp BMW.IV engine as an engine test-bed. The engine name plate had the old designation BMW.IIIa to obtain the flight permit from the Allied Monitoring Commission. *Leutnant* Franz Zeno Diemer reached a record altitude of 9,920 m on May 9th, 1919, and on June 17, 1919; however, the record was not recognized by the FAI at that time.

A postwar passenger conversion of the C.VII was also built. The resulting DFW P.I had an enclosed cabin for two passengers in place of the observer's open cockpit. Aircraft D 187 was the only P.I built. The Allied determination to destroy the German aviation industry as expressed in the terms of the Treaty of Versailles had the intended effect and DFW was soon put out of the aviation business.

Above & Below: The final DFW two-seater design is thought to have been assigned the military designation DFW C.VII but this is not confirmed. It was powered by a 220 hp Benz Bz.IVa. (Peter M. Bowers Collection/Museum of Flight)

Right: The DFW P.I was a postwar passenger conversion of the DFW C.VII with cabin for two passengers. Like most aircraft of the time, the pilot's cockpit was open so the pilot could feel the wind and hear the wires more clearly, important clues to aircraft performance given the limited instrumentation installed. D 187 shown was the only P.I built. (Peter M. Bowers Collection/Museum of Flight)

Euler C.I

August Euler as a noted aviation pioneer who designed and built aircraft and had a flying school. Twenty-nine Euler C.I aircraft, work numbers 204–231, were built during 1916 and used as operational trainers; the type does not appear in the Frontbeststand inventory. As a trainer, the type was delivered to *FEA* 10 at Böblinger, *FEA* 3 at Gotha, *FEA* 5 at Hannover, and *Kampf-Einsitzer-Staffel* 5 at Freiberg.

As was typical, the first generation C-Types were developed from B-types and fitted with more powerful engines to carry the additional weight of armement. The Euler C.I was developed from the 120 hp Euler B.I. Despite that it did not appear over the front, the Euler C.I was powered by the 160 hp Mercedes D.III that was in high demand to power single-seat and two-seat fighters. It was also armed with a fixed, synchronized gun for the pilot and a flexible gun for the observer.

Right: Euler C.I C3609/15 having a bad day. (P. M. Grosz Collection/STBD)

Above: Euler C I aircraft. (P. M. Grosz Collection/STBD)

Above: Euler C I with observer/gunner in front of the pilot.

Above: Euler C I two-seat reconnaissance aircraft in 1916.

Above: Euler C.I aircraft. (P. M. Grosz Collection/STBD)

Above: Euler Werke photo of an early Euler C.I with observer and gun in front cockpit. (P. M. Grosz Collection/STBD)

Above: An Euler C I, built by the Euler-Werke, C3626, Wnr. 225, 1916.

Above: Euler C I 3606, Wnr. 205, 1916.

DFW C.V 4918/16 of *Flieger-Abteilung (A)* 276.

DFW C.V(Av) C.287/18 of *Flieger-Abteilung (A)* 219, May 1918.

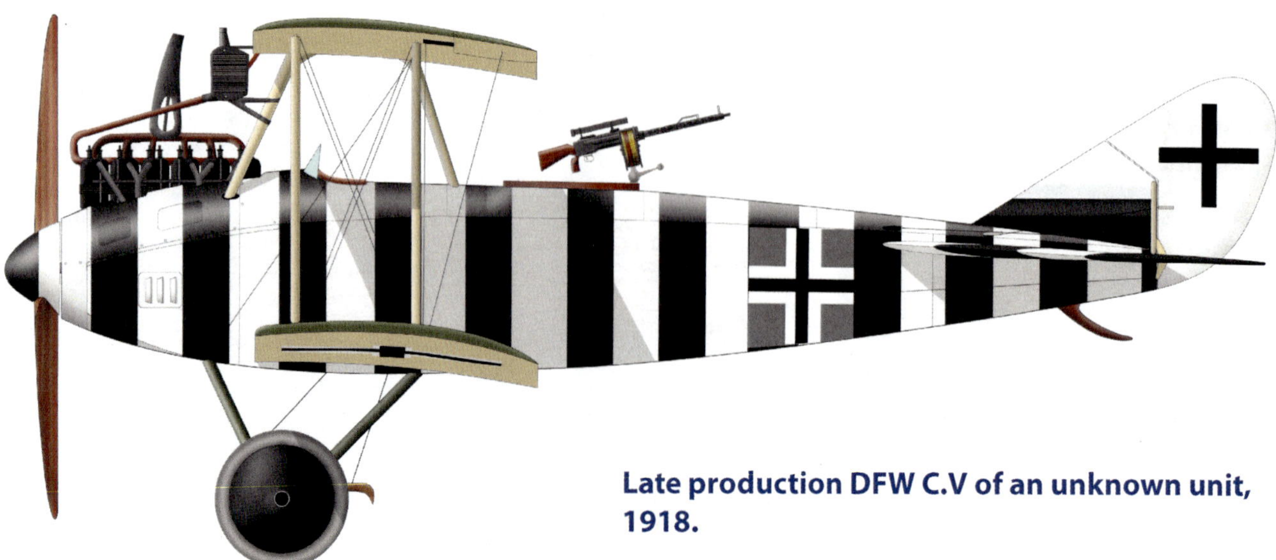

Late production DFW C.V of an unknown unit, 1918.

Made in the USA
Coppell, TX
07 June 2024

33243458R00105